This book belongs to

· ·

Thank you Duncan Lockerbie

Published by Julie MacLure, with the assistance of:

Lumphanan Press
Roddenbrae
Lumphanan
Aberdeenshire
AB31 4RN

http://www.lumphananpress.co.uk

ISBN: 978-0-9955323-9-7

Interactive Learning Psychological Services Limited

Progressive MATHS

for School

Julie MacLure

CONTENTS

Unit 1

Counting

Teacher Note

Counting

Numbers to 10:
recognition of digit and word

Outcome: Number, money and measurement

Strand: Add and subtract

Target: add and subtract – mentally for numbers 0 to 10.

Aims

The pupils should know the numbers up to 10 and be able to recognise a number as a digit and in words.

Objectives

To teach the relationship between number, digit and written word.

Introduction

Practice materials such as blocks should be used to count first. Pupils should be encouraged to count out loud.

Counting

Colour 1

Copy and complete

one

o ....

Colour 2

Copy and complete

two

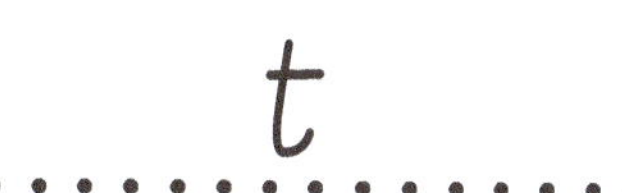
t ....

Colour 3

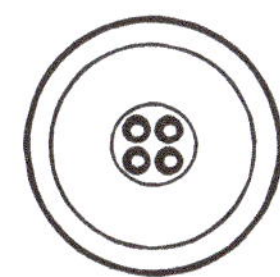
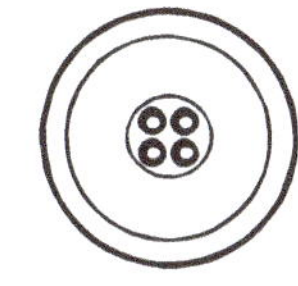

Copy and complete

three

t ....

Counting

Colour 4 write

four

....

Colour 5 write

five

....

Colour 6 write

six

....

Counting

Colour 7　　　　write

seven

....

Colour 8　　　　write

eight

....

Colour 9　　　　write

nine

....

Counting

Colour one   **write**

one, 1

....

Colour two   **write**

two, 2

....

Colour three   **write**

three, 3

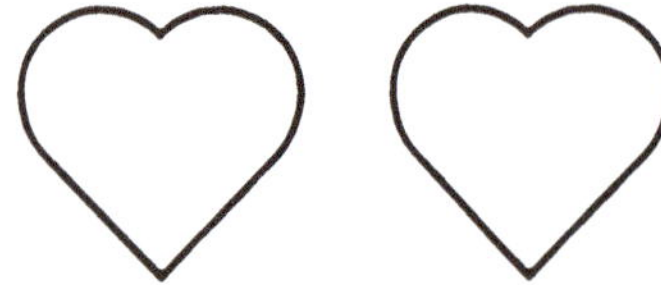

....

Counting

Colour four
write
four, 4
Colour five
write
five, 5
Colour six
write
six, 6

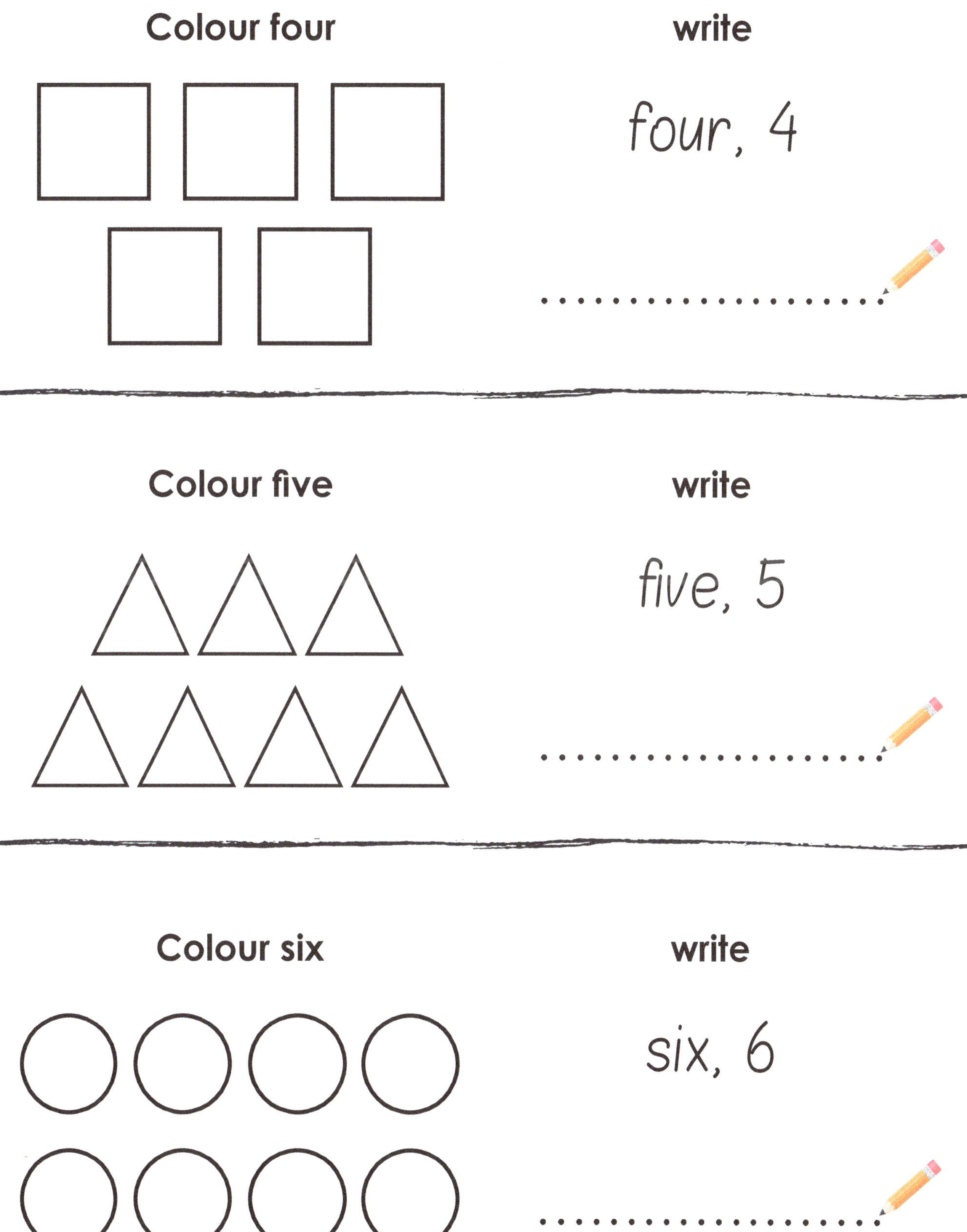

Counting

Draw and colour seven   **write**

seven, 7

....

Draw and colour eight   **write**

eight, 8

....

Draw and colour nine   **write**

nine, 9

....

Matching

Match

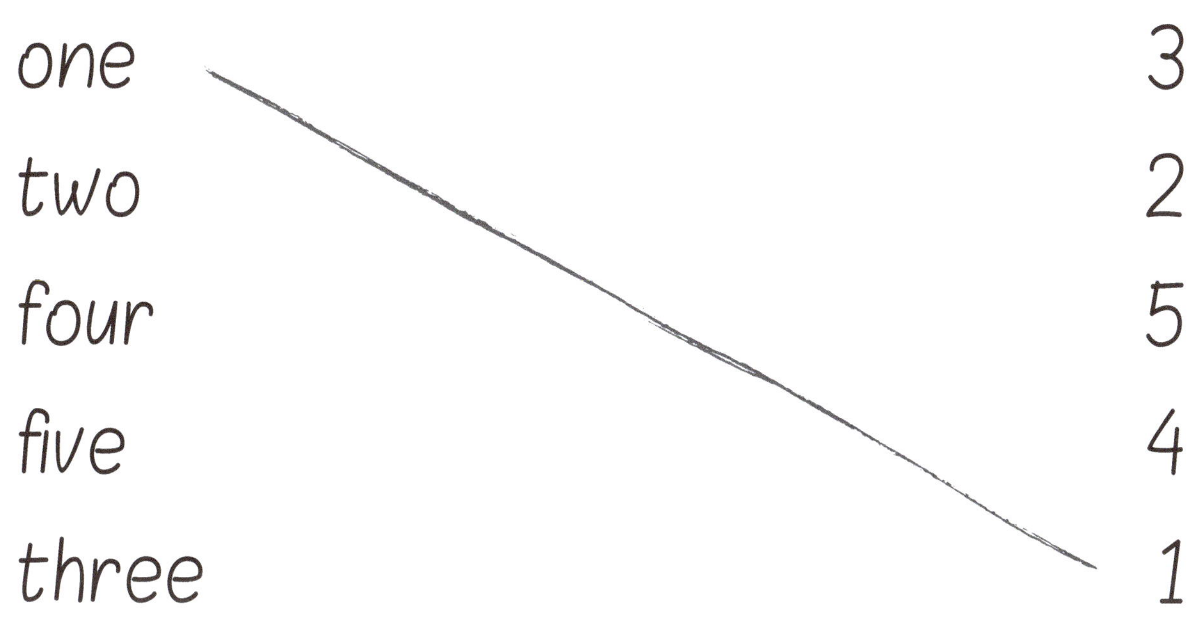

one	3
two	2
four	5
five	4
three	1

Count, then match

| 4 |
| 1 |
| 2 |
| 3 |
| 5 |

Matching

Count, then match

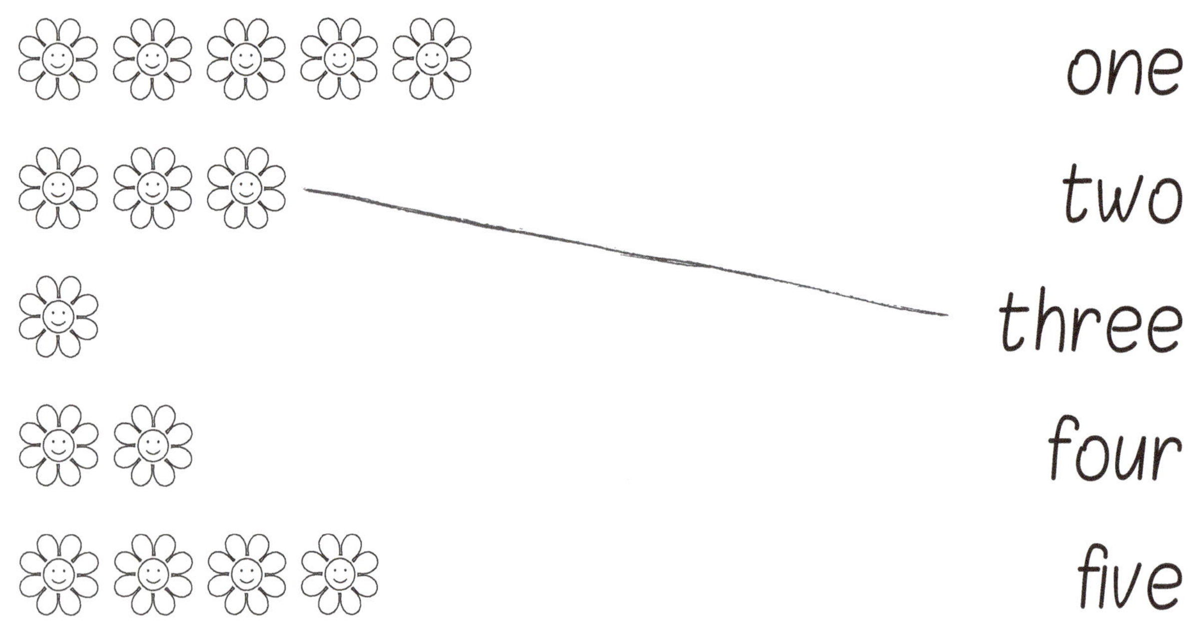

Match

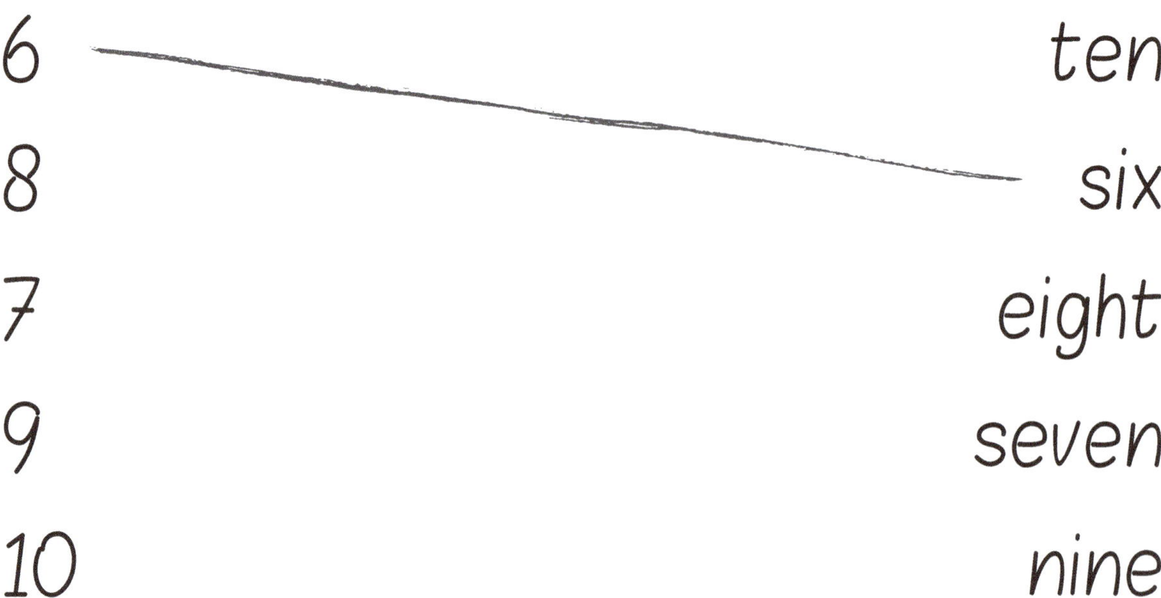

Matching

Count, then match

9

10

8

7

6

Count, then match

6

7

8

9

10

Matching

12

Put the numbers in order

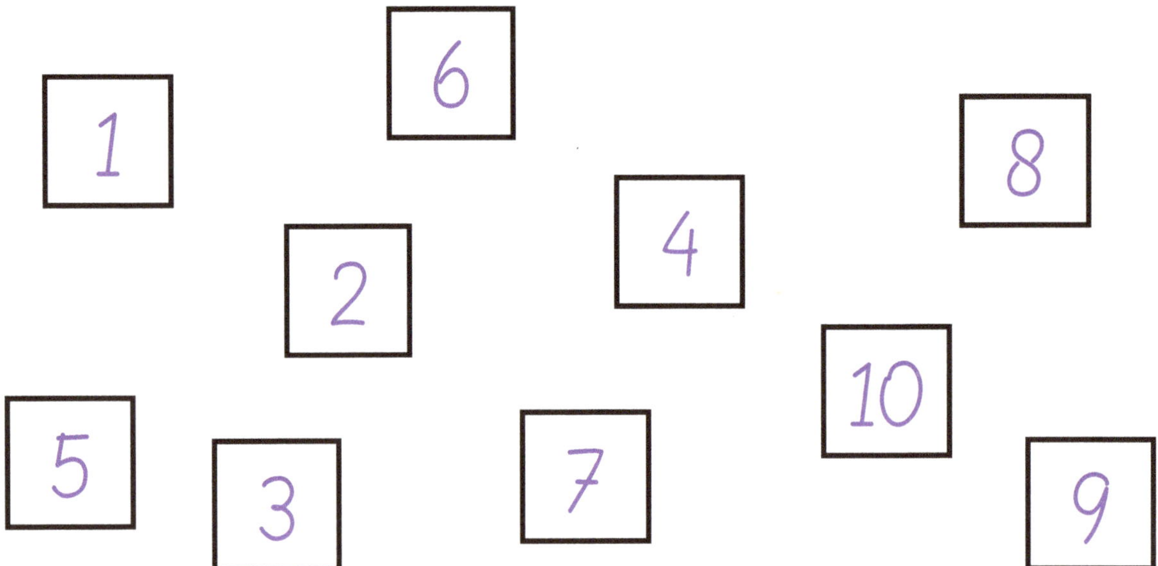

Join the dots from 0 to 6 and colour

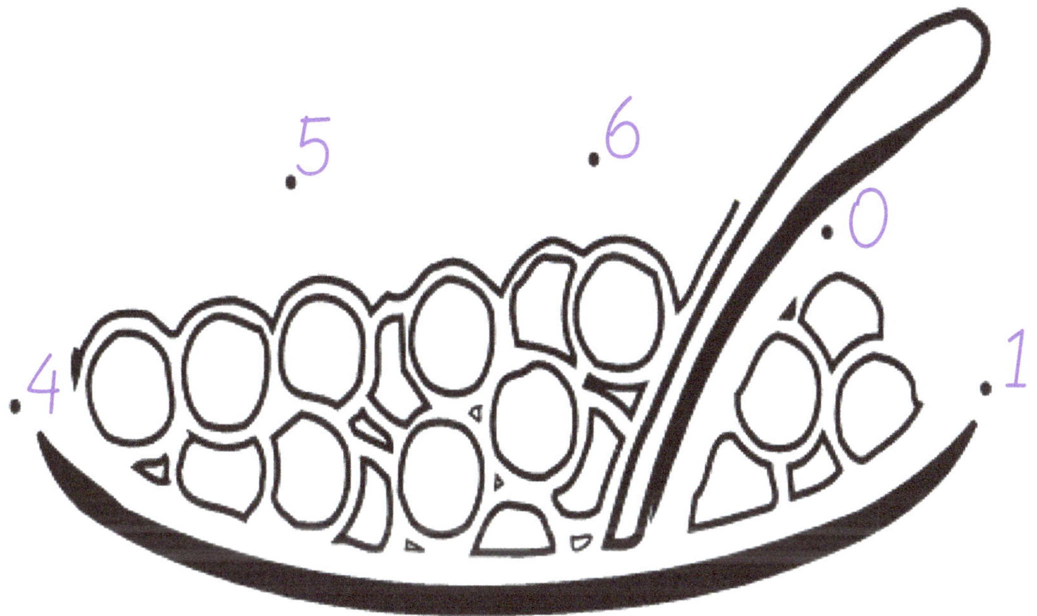

Join the dots from 0 to 8 and colour

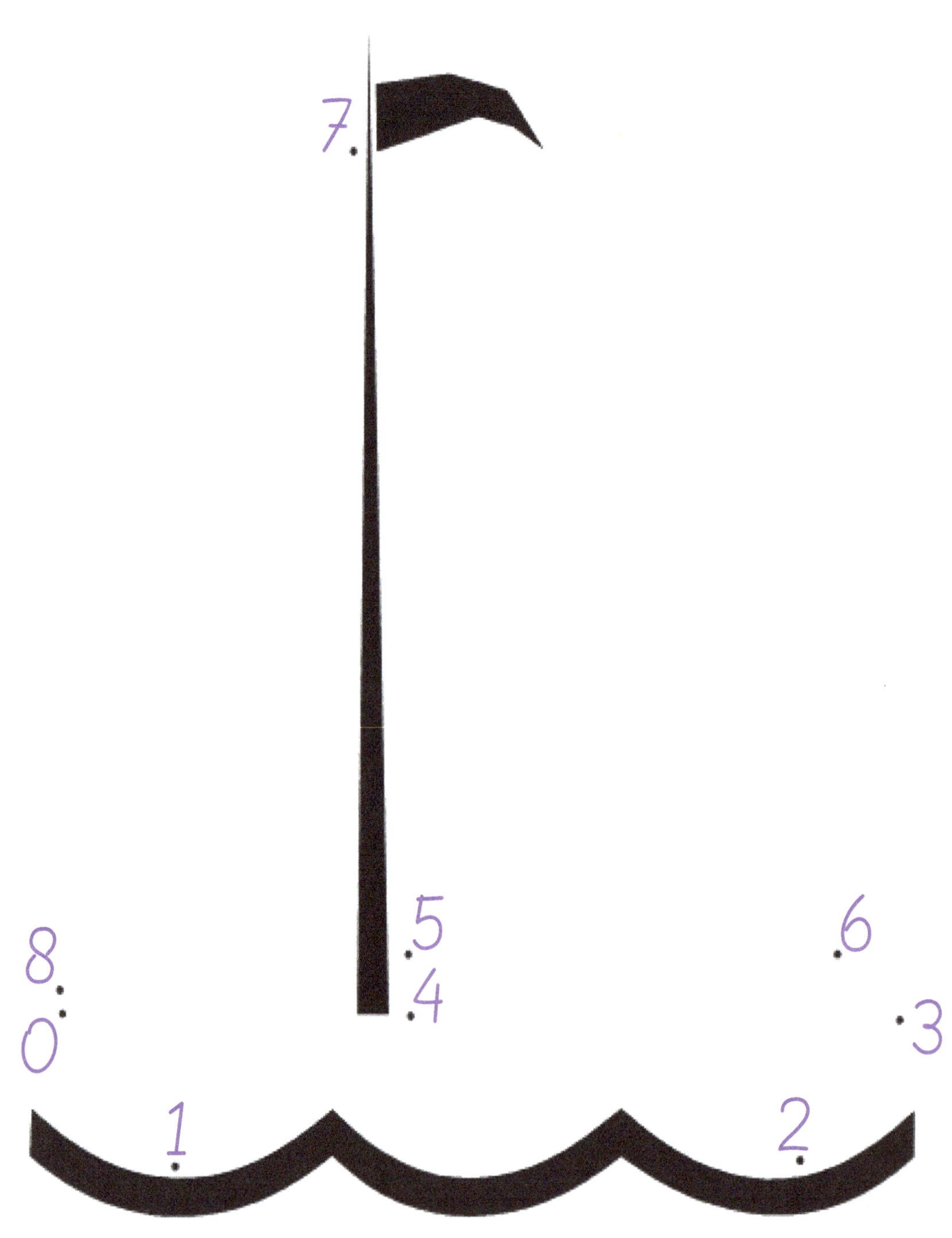

Notes

15

Notes

16

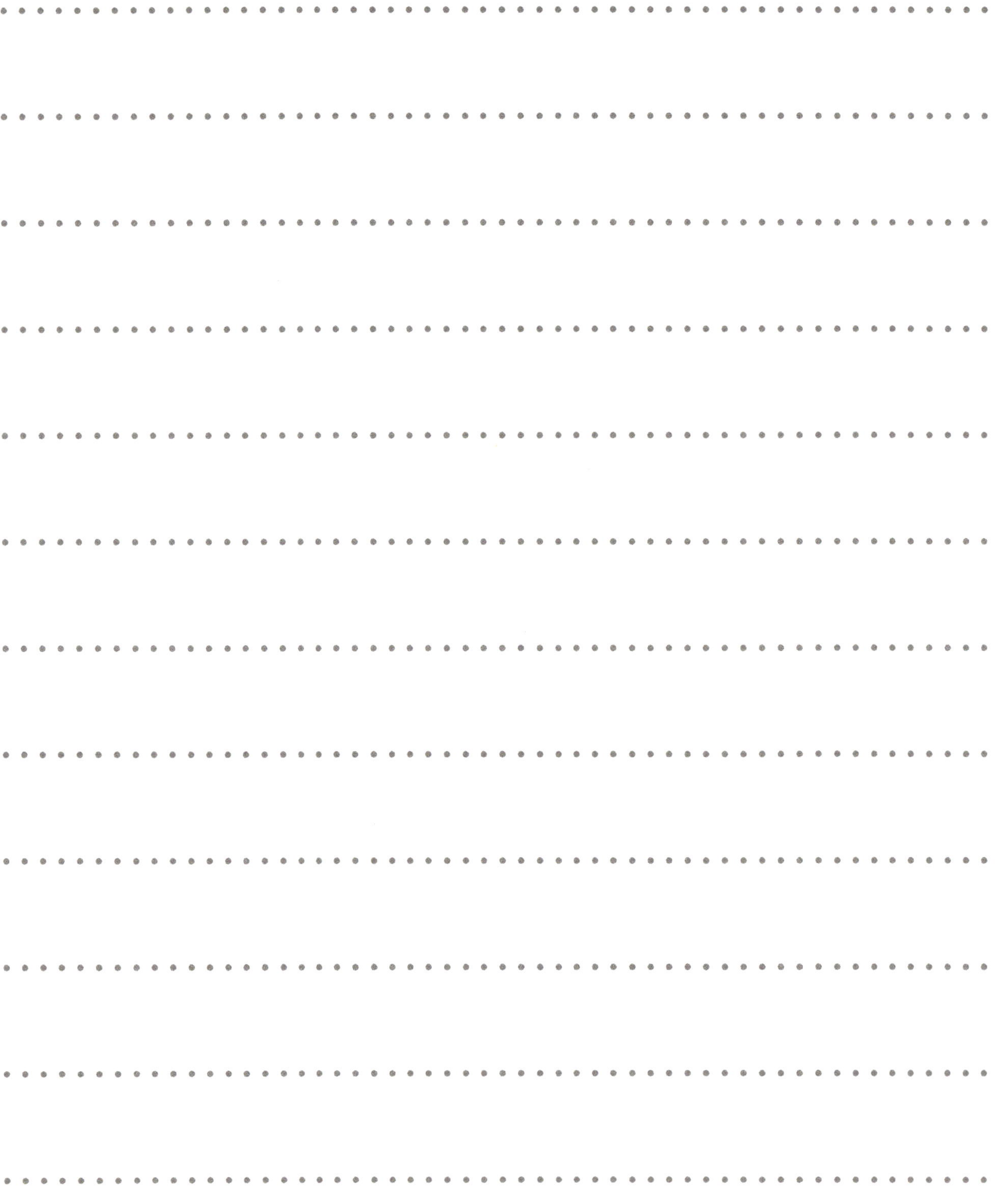

Problem solving

For this activity you will need, a paper fastener and a scissors.

Cut around the clock face after colouring it in. Cut out the clock hands. Then attach to the centre of the clock with a paper fastener.

Group Activity

Play guess the time. Set the hands to an on the hour time then ask your group if they know the answer. The group then write the time down. See how many times you get right. Each take a turn at setting a time.

Unit 2

Adding

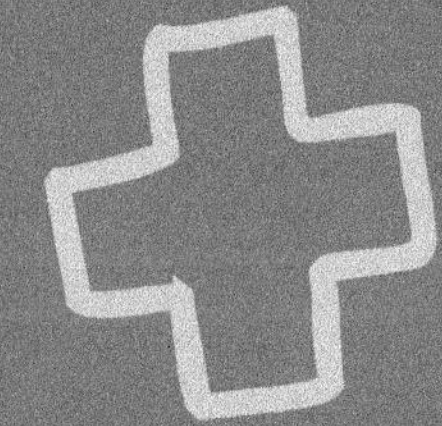
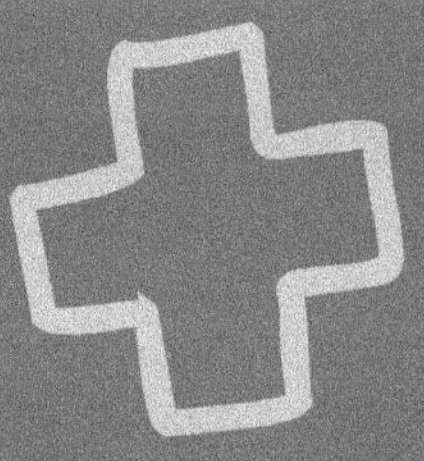

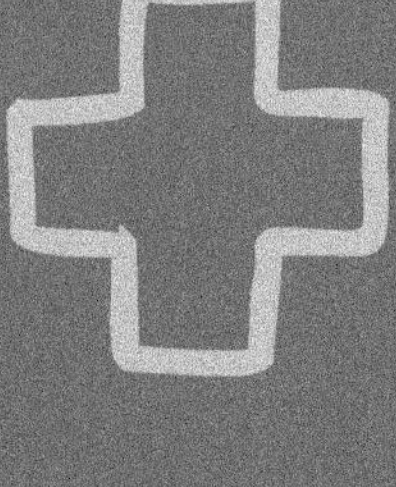

Teacher Note

Adding

Numbers to 10:
recognition of digit and word

Outcome: Number, money and measurement

Strand: Add and subtract

Target: add and subtract – mentally for numbers 0 to 10.

Level: A

Aims

The pupils should know how to add.

Objectives

To reinforce number bonds up to 10.

Introduction

Practice materials should be used before and along with the worksheets, if necessary.

How many?

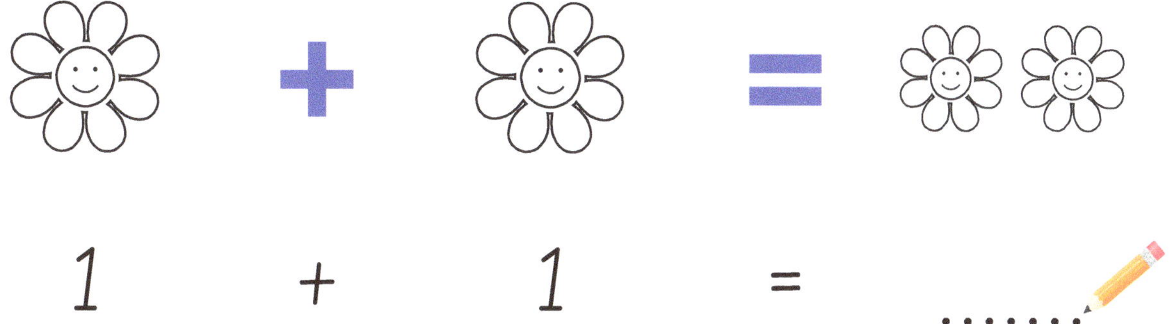

1 + 1 = ...

complete

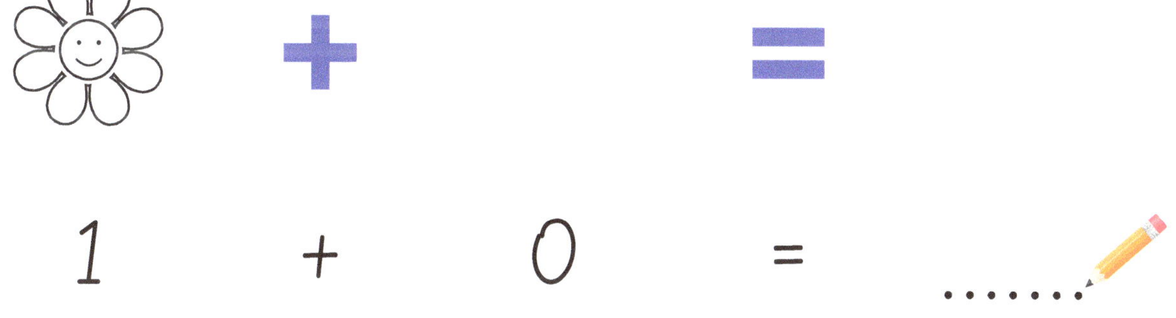

1 + 0 = ...

complete

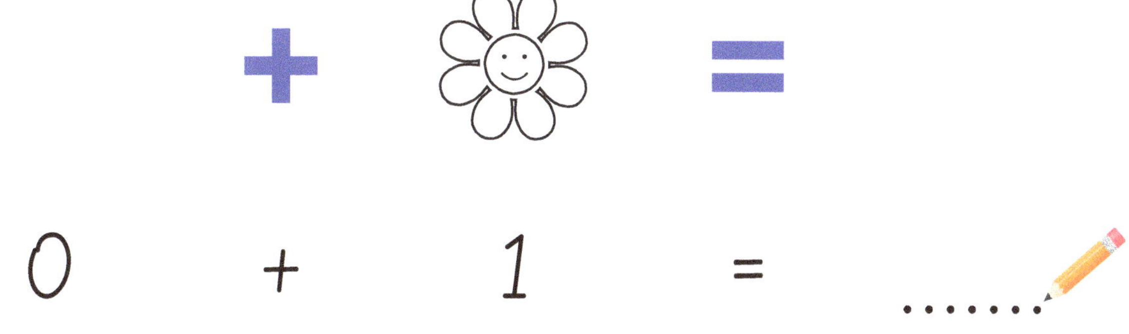

0 + 1 = ...

How many?

2

2 + 0 = ...

complete

1 + 1 = ...

complete

0 + 2 = ...

How many?

complete

3   +   0   = ...

---

complete

2   +   1   = ...

---

complete

1   +   2   = ...

How many?

2

complete

$0 \quad + \quad 3 \quad = \quad \ldots$

complete

$0 \quad + \quad 4 \quad = \quad \ldots$

complete

$1 \quad + \quad 3 \quad = \quad \ldots$

How many?

complete

2 + 2 = ...

complete

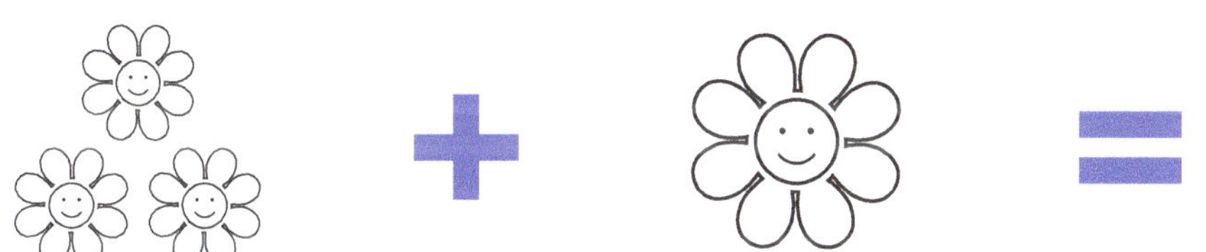

3 + 1 = ...

complete

4 + 0 = ...

How many?

complete

0   +   5   = ...

---

complete

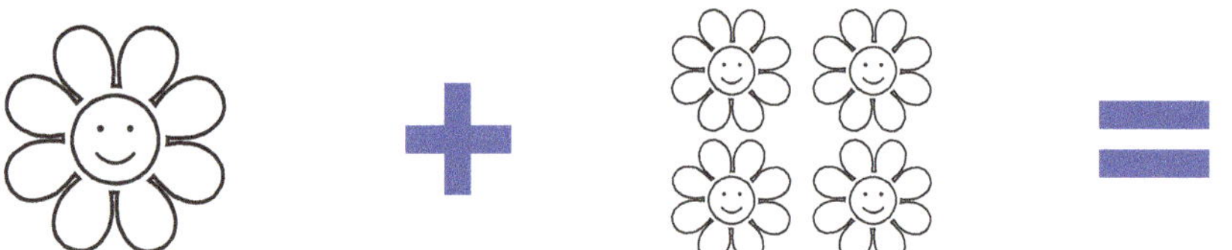

1   +   4   = ...

---

complete

2   +   3   = ...

How many?

complete

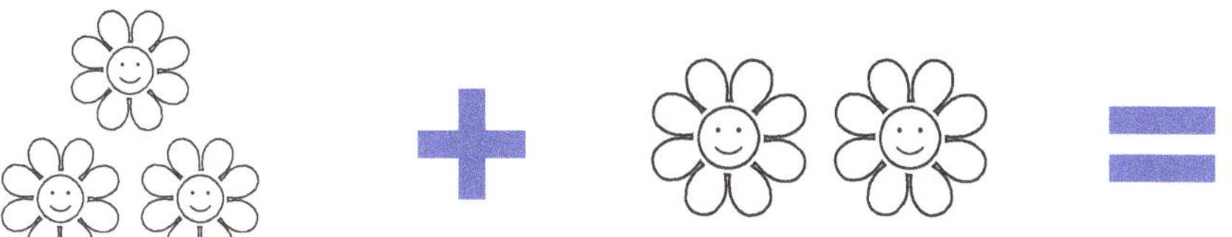

$$3 \quad + \quad 2 \quad = \quad \dots\dots$$

complete

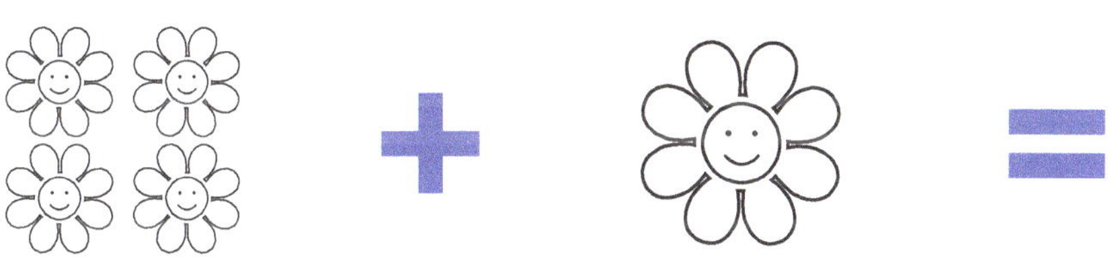

$$4 \quad + \quad 1 \quad = \quad \dots\dots$$

complete

$$5 \quad + \quad 0 \quad = \quad \dots\dots$$

How many?

complete

$$0 \quad + \quad 6 \quad = \quad \dots$$

complete

$$1 \quad + \quad 5 \quad = \quad \dots$$

complete

$$2 \quad + \quad 4 \quad = \quad \dots$$

How many?

complete

3   +   3   = ...

complete

4   +   2   = ...

complete

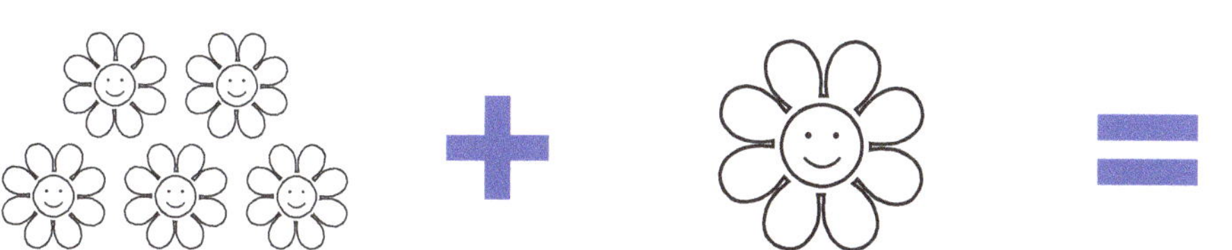

5   +   1   = ...

How many?

complete

6 + 0 = ...

---

complete

0 + 7 = ...

---

complete

1 + 6 = ...

How many?

complete

2   +   5   = ...

---

complete

3   +   4   = ...

---

complete

4   +   3   = ...

How many?

complete

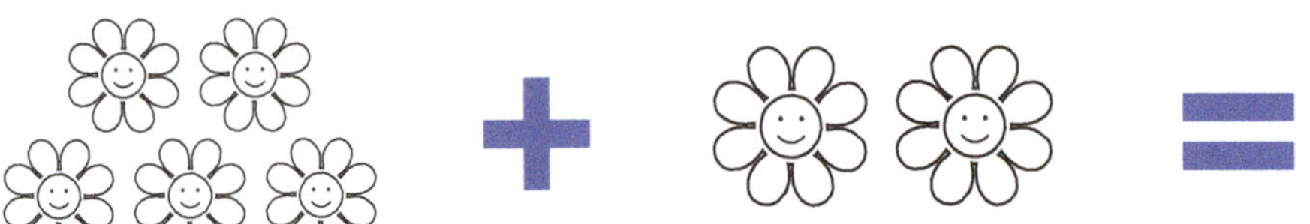

5   +   2   = ...

---

complete

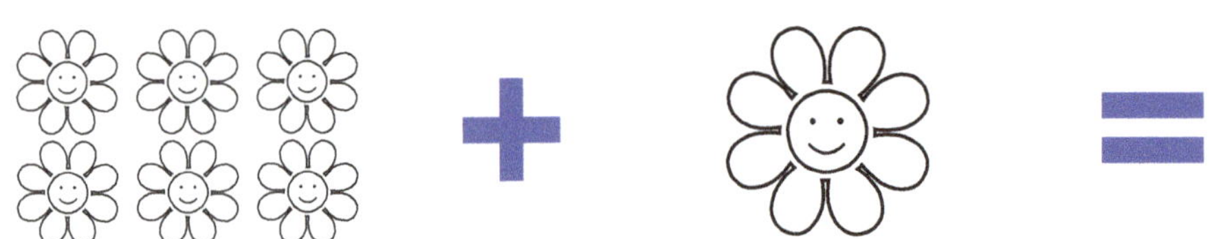

6   +   1   = ...

---

complete

7   +   0   = ...

How many?

complete

$+$  $=$

0   +   8   = ...

complete

1   +   7   = ...

complete

2   +   6   = ...

How many?

complete

3   +   5   = ...

complete

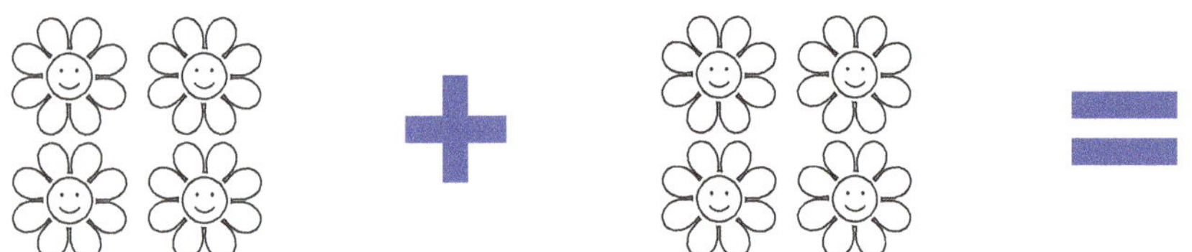

4   +   4   = ...

complete

5   +   3   = ...

How many?

complete

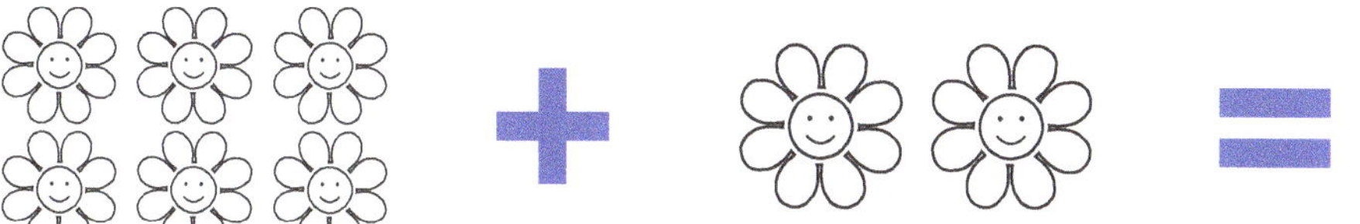

6 + 2 = ...

complete

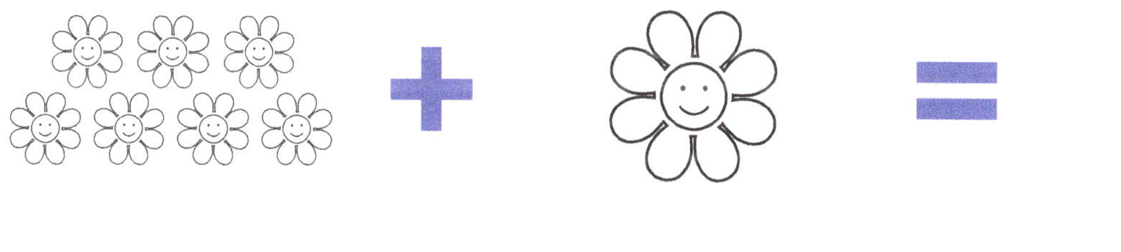

7 + 1 = ...

complete

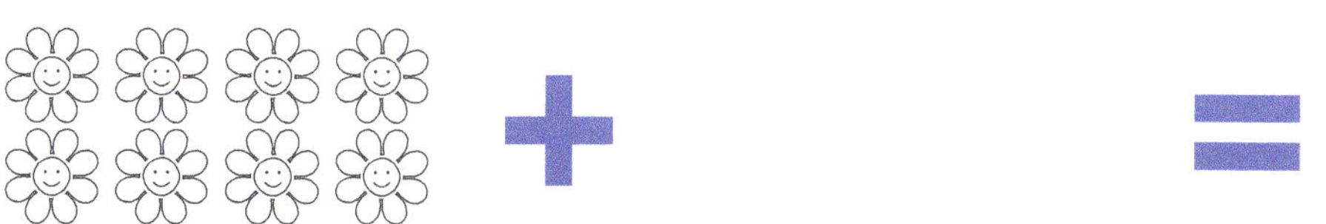

8 + 0 = ...

How many?

complete

$$0 \quad + \quad 9 \quad = \quad \text{........}$$

---

complete

$$1 \quad + \quad 8 \quad = \quad \text{........}$$

---

complete

$$2 \quad + \quad 7 \quad = \quad \text{........}$$

How many?

complete

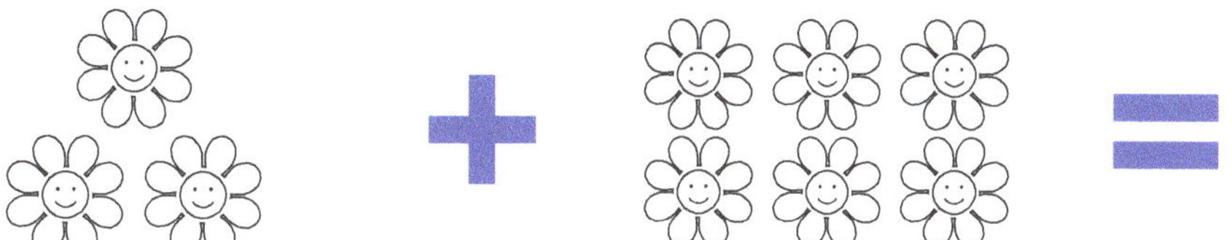

3   +   6   = ...

complete

4   +   5   = ...

complete

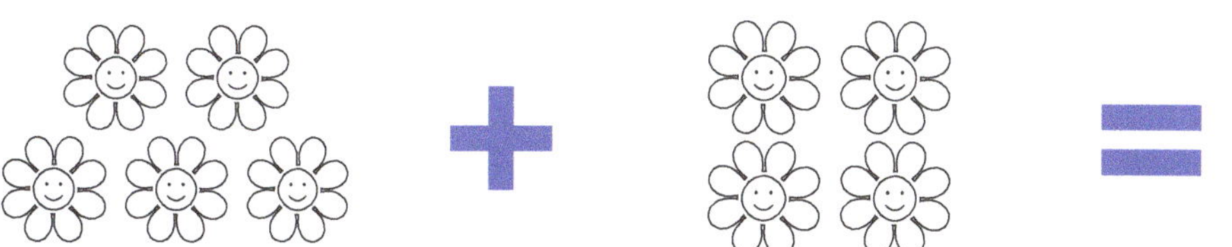

5   +   4   = ...

How many?

complete

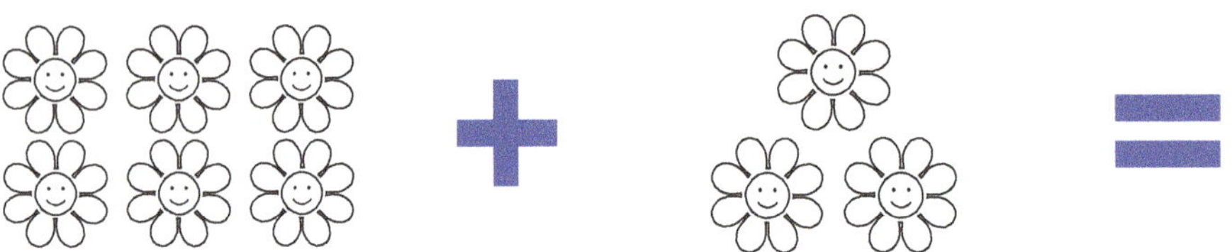

6   +   3   = ...

complete

7   +   2   = ...

complete

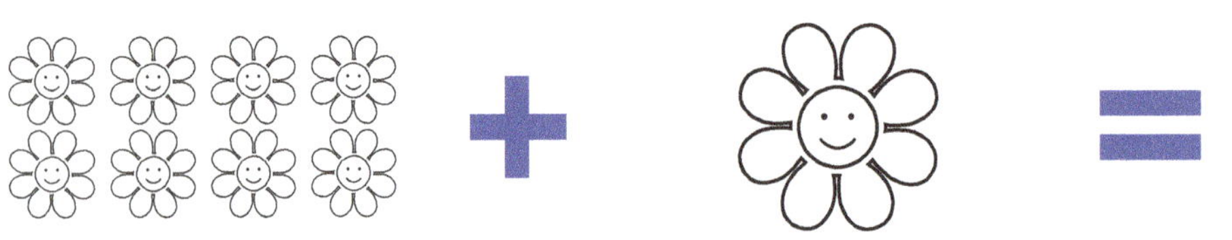

8   +   1   = ...

How many?

complete

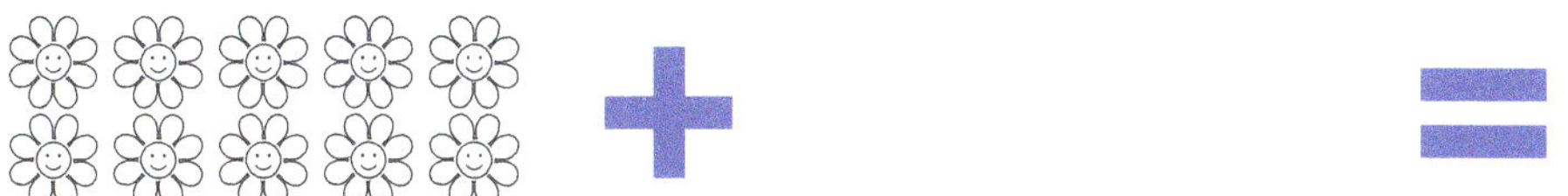

$$10 \quad + \quad 0 \quad = \quad \dots\dots$$

complete

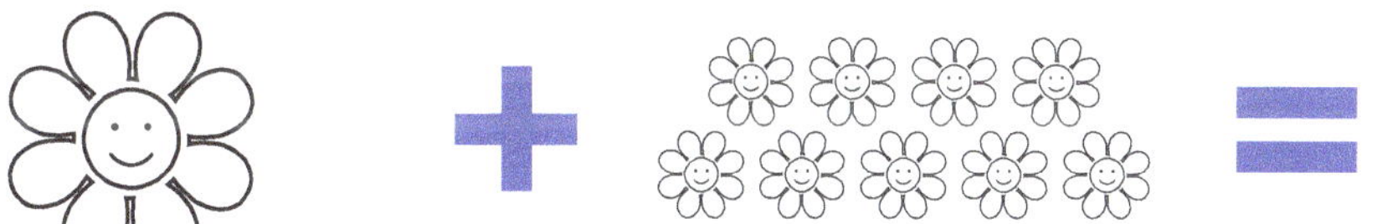

$$1 \quad + \quad 9 \quad = \quad \dots\dots$$

complete

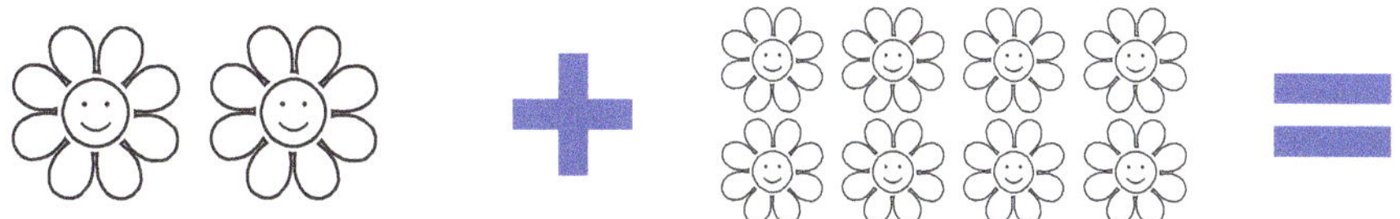

$$2 \quad + \quad 8 \quad = \quad \dots\dots$$

Notes

Notes

41

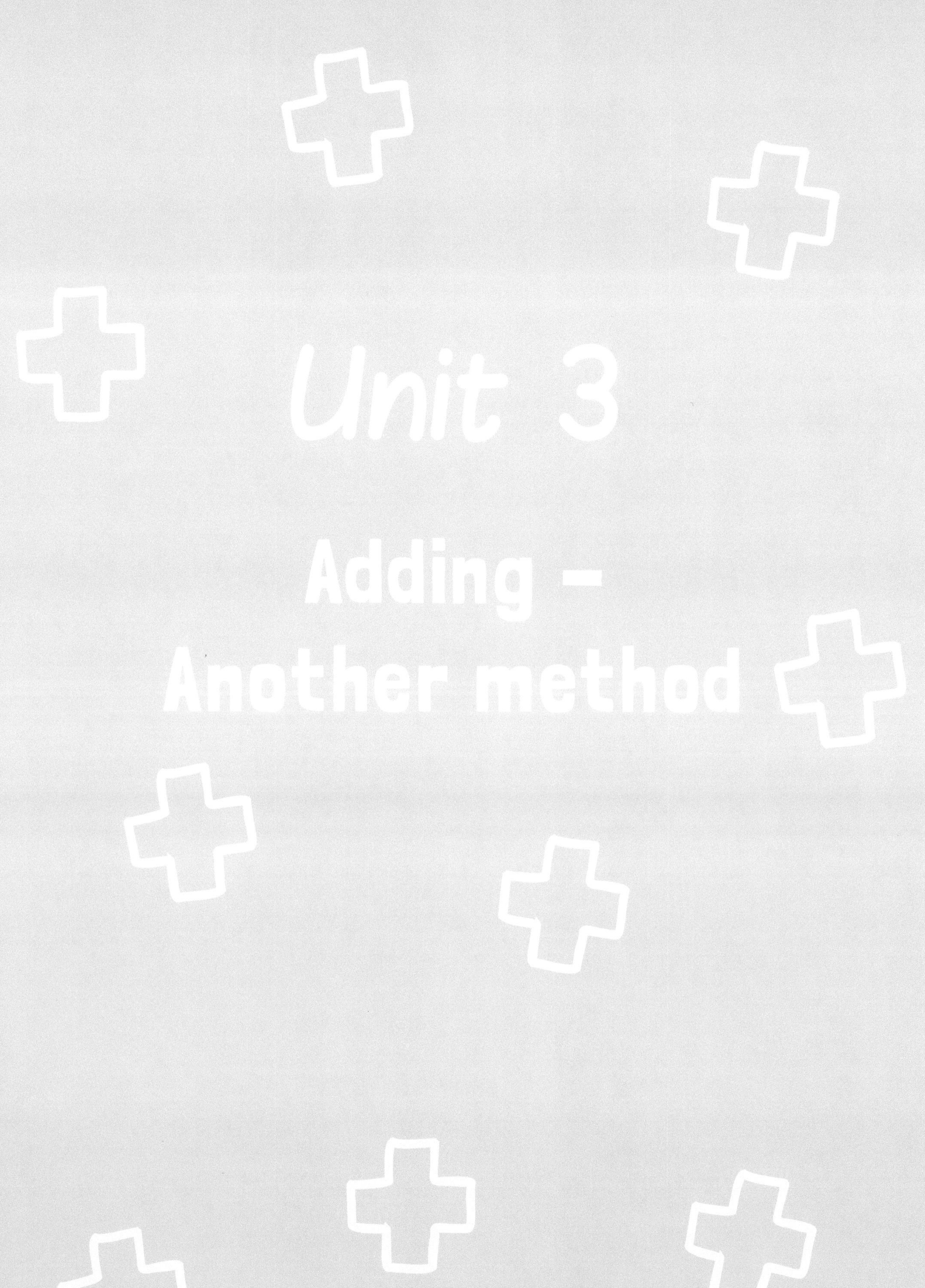

Unit 3

Adding –
Another method

Teacher Note

Adding – Another method

Numbers to 20:
recognition of digit and word

Outcome: Number, money and measurement

Strand: Add and subtract

Target: add and subtract – mentally for numbers 0 to 20.

Level: A

Aims

To count up to 20.

Objectives

To recognise, put in order and add whole numbers up to 20.

Activities

The activities are designed for self learning. An abacus can be used to develop these skills.

Count

Colour   **Write**

1 bundle of 10 plus 1

1 bundle of 10 plus 2

1 bundle of 10 plus 3

14

15

16

17

18

19

20

Count

3

Tens	Units		
1	1	*11*	*Eleven*
1	2	*12*	*Twelve*
1	3		
1	4		

Count

1 + 11 =

1 + 12 =

One more than 11 =

One more than 12 =

Two more than 11 =

Notes

Notes

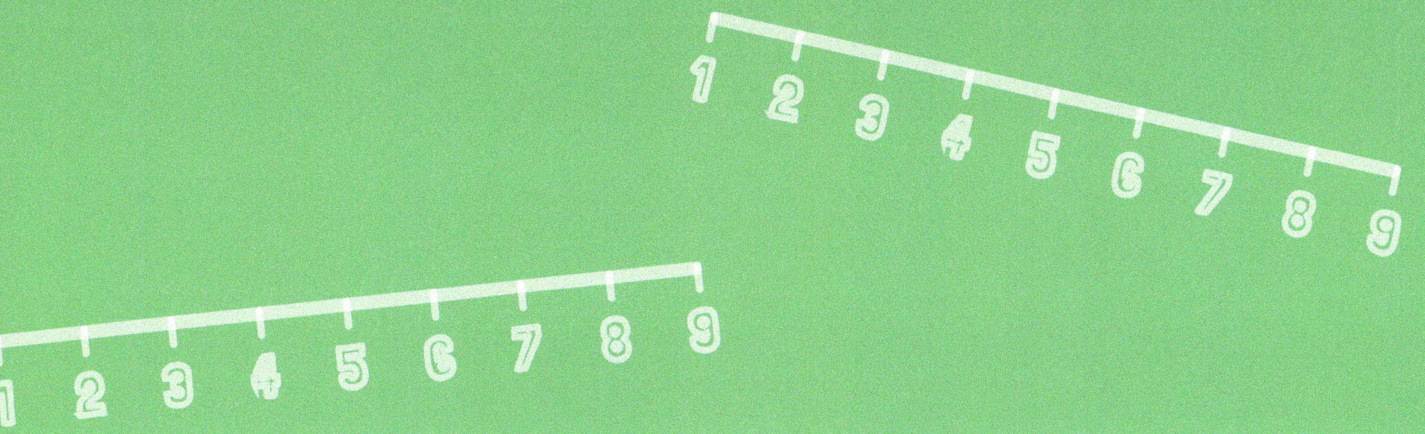

Unit 4

Number line

Teacher Note

Number line

Reinforcing number bonds to 20:
use of number line

Outcome: Number, money and measurement

Strand: Add and subtract

Level: A – add and subtract – mentally for numbers 0 to 10.
B – add and subtract – mentally for numbers 0 to 20.

Aims

The pupils should be able to 'add on' up to 20.

Objectives

To reinforce number bonds up to 20.

Introduction

The pupils should be introduced to the number line. Number lines could be made from card and one given to each pupil. Pupils should be instructed to add in 'jumps' forward on the number line. The pupils should be encouraged to count out loud and then complete the sum.

Draw in jumps forward

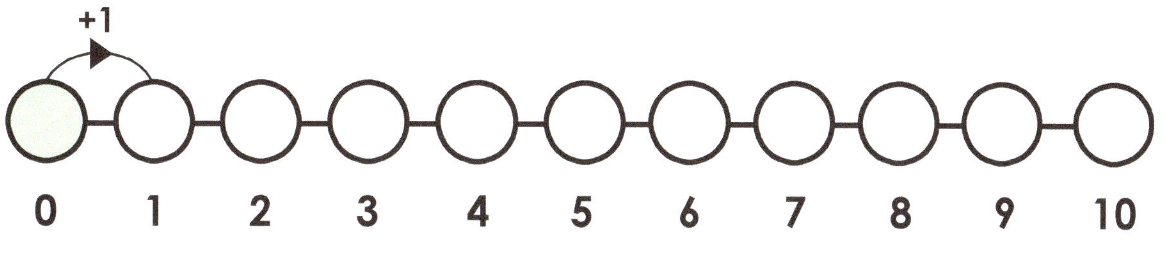

Start at 0, add 1

$$0 \quad + \quad 1 \quad = \quad 1$$

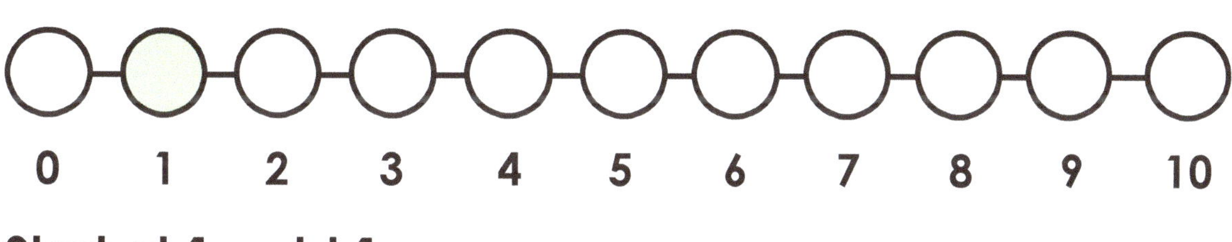

Start at 1, add 1

$$1 \quad + \quad 1 \quad = \quad \underline{\quad\quad}$$

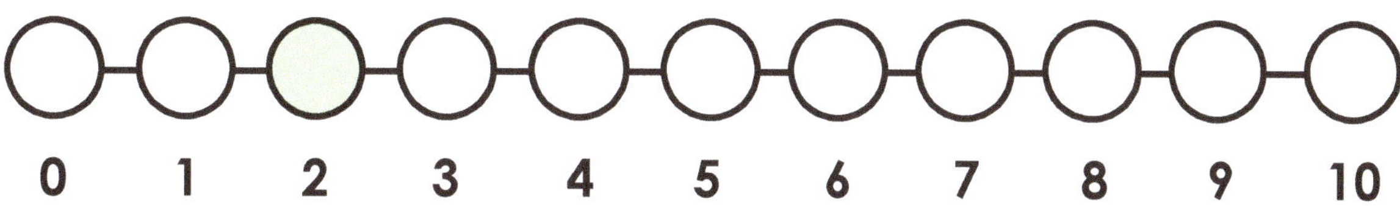

Start at 2, add 1

$$2 \quad + \quad 1 \quad = \quad \underline{\quad\quad}$$

Draw in jumps forward

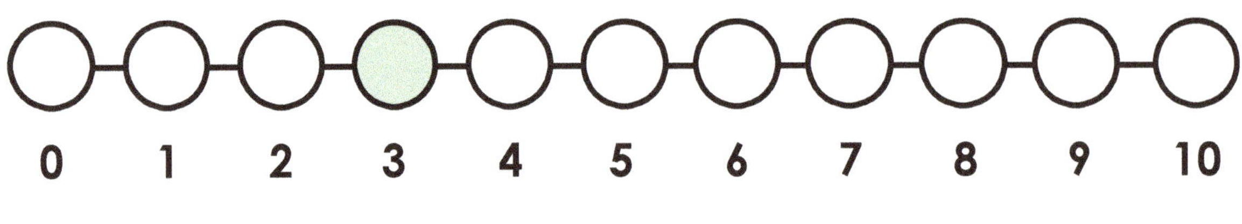

Start at 3, add 1

$$3 \quad + \quad 1 \quad = \quad \underline{\hspace{2cm}}$$

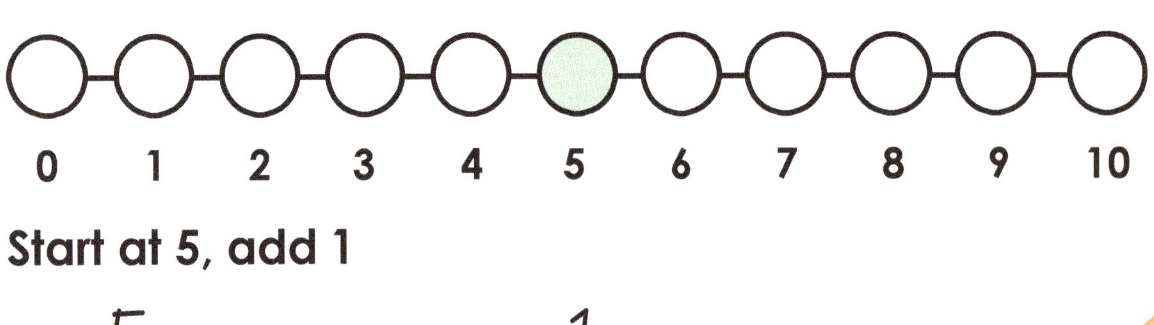

Start at 4, add 1

$$4 \quad + \quad 1 \quad = \quad \underline{\hspace{2cm}}$$

Start at 5, add 1

$$5 \quad + \quad 1 \quad = \quad \underline{\hspace{2cm}}$$

Draw in jumps forward

0　1　2　3　4　5　6　7　8　9　10

Start at 6, add 1

6　　　+　　　1　　　=　　......

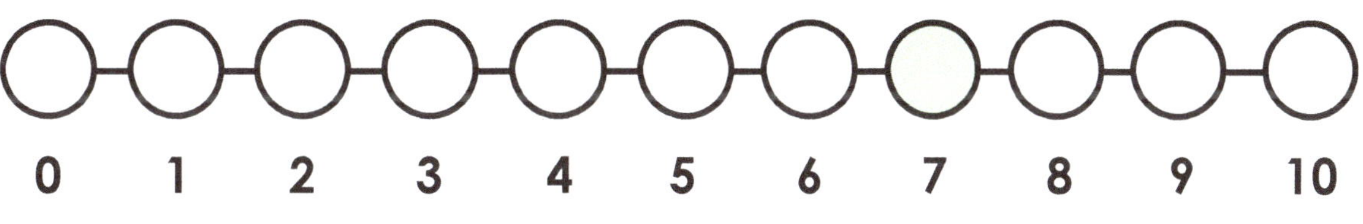

0　1　2　3　4　5　6　7　8　9　10

Start at 7, add 1

7　　　+　　　1　　　=　　......

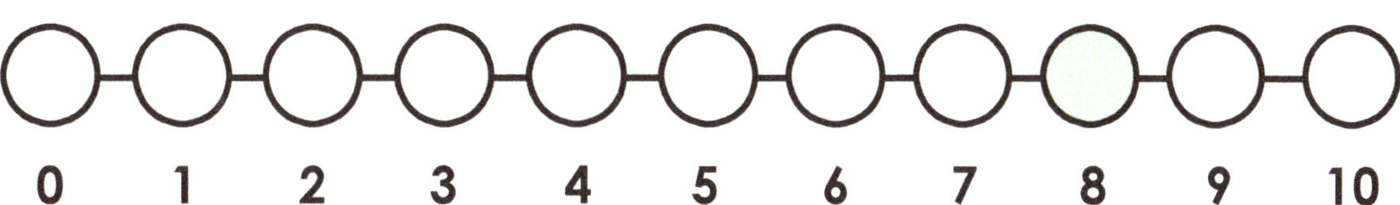

0　1　2　3　4　5　6　7　8　9　10

Start at 8, add 1

8　　　+　　　1　　　=　　......

Draw in jumps forward

Start at 9, add 1

9   +   1   = ...

Start at 0, add 2

0   +   2   = ...

Start at 1, add 2

1   +   2   = ...

Draw in jumps forward

Start at 2, add 2

$$2 \quad + \quad 2 \quad = \quad \text{......}$$

Start at 3, add 2

$$3 \quad + \quad 2 \quad = \quad \text{......}$$

Start at 4, add 2

$$4 \quad + \quad 2 \quad = \quad \text{......}$$

Draw in jumps forward

4

0 1 2 3 4 5 6 7 8 9 10 11

Start at 5, add 2

$$5 + 2 = \text{\_\_\_\_\_\_}$$

0 1 2 3 4 5 6 7 8 9 10 11

Start at 6, add 2

$$6 + 2 = \text{\_\_\_\_\_\_}$$

0 1 2 3 4 5 6 7 8 9 10 11

Start at 7, add 2

$$7 + 2 = \text{\_\_\_\_\_\_}$$

Draw in jumps forward

0 1 2 3 4 5 6 7 8 9 10 11

Start at 8, add 2

$$8 \quad + \quad 2 \quad = \quad \underline{\quad\quad}$$

0 1 2 3 4 5 6 7 8 9 10 11

Start at 9, add 2

$$9 \quad + \quad 2 \quad = \quad \underline{\quad\quad}$$

0 1 2 3 4 5 6 7 8 9 10 11 12

Start at 0, add 3

$$0 \quad + \quad 3 \quad = \quad \underline{\quad\quad}$$

Draw in jumps forward

Start at 1, add 3

$$1 \quad + \quad 3 \quad = \quad \underline{}$$

Start at 2, add 3

$$2 \quad + \quad 3 \quad = \quad \underline{}$$

Start at 3, add 3

$$3 \quad + \quad 3 \quad = \quad \underline{}$$

Draw in jumps forward

Start at 4, add 3

$$4 \quad + \quad 3 \quad = \quad \text{..............}$$

Start at 5, add 3

$$5 \quad + \quad 3 \quad = \quad \text{..............}$$

Start at 6, add 3

$$6 \quad + \quad 3 \quad = \quad \text{..............}$$

Draw in jumps forward

4

0　1　2　3　4　5　6　7　8　9　10　11　12

Start at 7, add 3

7　　+　　3　　=　　..............

0　1　2　3　4　5　6　7　8　9　10　11　12

Start at 8, add 3

8　　+　　3　　=　　..............

0　1　2　3　4　5　6　7　8　9　10　11　12

Start at 9, add 3

9　　+　　3　　=　　..............

Draw in jumps forward

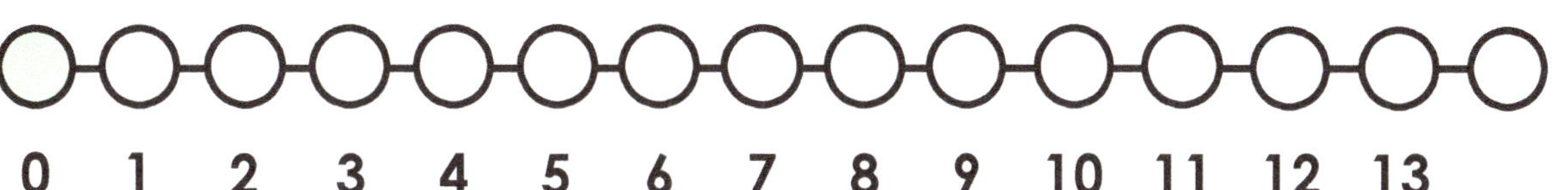

0 1 2 3 4 5 6 7 8 9 10 11 12 13

Start at 0, add 4

$$0 \quad + \quad 4 \quad = \quad \underline{\hspace{2cm}}$$

0 1 2 3 4 5 6 7 8 9 10 11

Start at 1, add 4

$$1 \quad + \quad 4 \quad = \quad \underline{\hspace{2cm}}$$

0 1 2 3 4 5 6 7 8 9 10 11 12

Start at 2, add 4

$$2 \quad + \quad 4 \quad = \quad \underline{\hspace{2cm}}$$

Draw in jumps forward

0 1 2 3 4 5 6 7 8 9 10 11 12 13

Start at 3, add 4

3 + 4 = ...

0 1 2 3 4 5 6 7 8 9 10 11

Start at 4, add 4

4 + 4 = ...

0 1 2 3 4 5 6 7 8 9 10 11 12

Start at 5, add 4

5 + 4 = ...

Draw in jumps forward

Start at 6, add 4

$$6 \quad + \quad 4 \quad = \quad \underline{\hspace{2cm}}$$

Start at 7, add 4

$$7 \quad + \quad 4 \quad = \quad \underline{\hspace{2cm}}$$

Start at 8, add 4

$$8 \quad + \quad 4 \quad = \quad \underline{\hspace{2cm}}$$

Draw in jumps forward

Start at 9, add 4

$$9 \quad + \quad 4 \quad = \quad \text{.........}$$

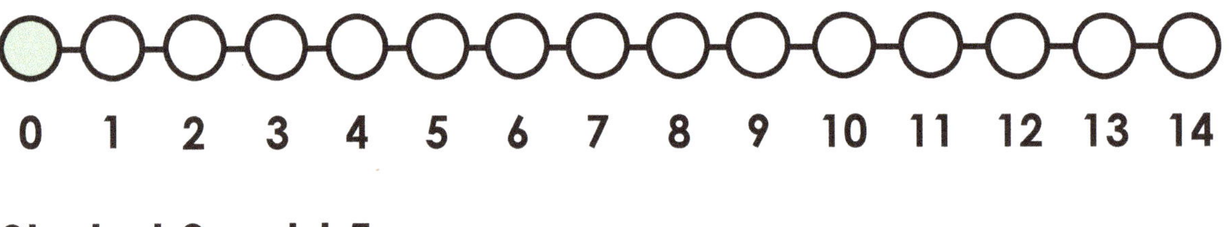

Start at 0, add 5

$$0 \quad + \quad 5 \quad = \quad \text{.........}$$

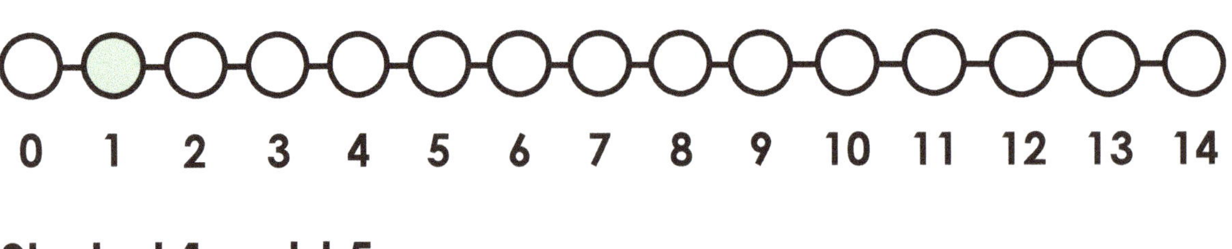

Start at 1, add 5

$$1 \quad + \quad 5 \quad = \quad \text{.........}$$

Draw in jumps forward

0 1 2 3 4 5 6 7 8 9 10 11 12 13 14

Start at 2, add 5

2 + 5 = 

0 1 2 3 4 5 6 7 8 9 10 11 12 13 14

Start at 3, add 5

3 + 5 = 

0 1 2 3 4 5 6 7 8 9 10 11 12 13 14

Start at 4, add 5

4 + 5 =

Draw in jumps forward

4

0 1 2 3 4 5 6 7 8 9 10 11 12 13 14

Start at 5, add 5

$$5 \quad + \quad 5 \quad = \quad \text{............}$$

0 1 2 3 4 5 6 7 8 9 10 11 12 13 14

Start at 6, add 5

$$6 \quad + \quad 5 \quad = \quad \text{............}$$

0 1 2 3 4 5 6 7 8 9 10 11 12 13 14

Start at 7, add 5

$$7 \quad + \quad 5 \quad = \quad \text{............}$$

Draw in jumps forward

Start at 8, add 5

8   +   5   = ...

Start at 9, add 5

9   +   5   = ...

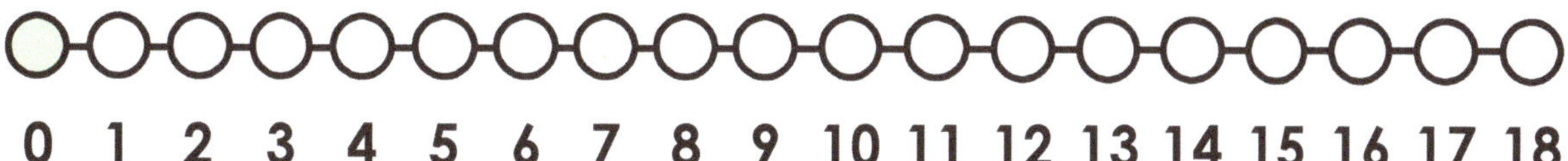

Start at 0, add 6

0   +   6   = ...

Draw in jumps forward

4

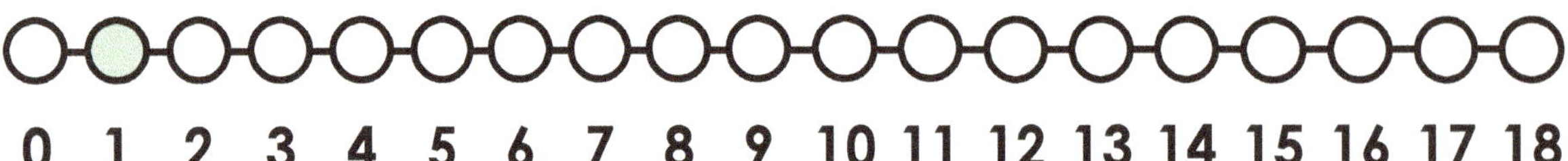

0 1 2 3 4 5 6 7 8 9 10 11 12 13 14 15 16 17 18

Start at 1, add 6

1 + 6 = ...

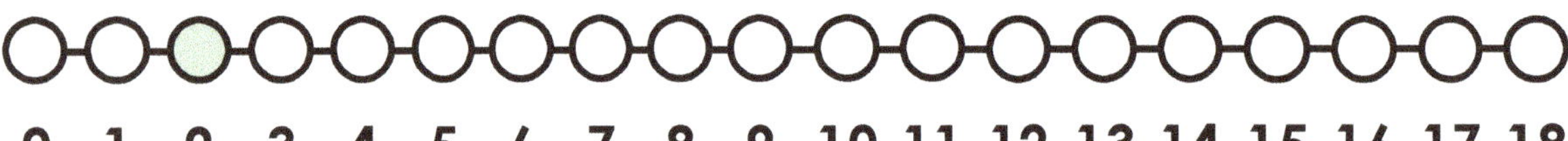

0 1 2 3 4 5 6 7 8 9 10 11 12 13 14 15 16 17 18

Start at 2, add 6

2 + 6 = ...

0 1 2 3 4 5 6 7 8 9 10 11 12 13 14 15 16 17 18

Start at 3, add 6

3 + 6 = ...

Draw in jumps forward

Start at 4, add 6

4   +   6   = ...

Start at 5, add 6

5   +   6   = ...

Start at 6, add 6

6   +   6   = ...

Draw in jumps forward

0 1 2 3 4 5 6 7 8 9 10 11 12 13 14 15 16 17 18

Start at 7, add 6

7 + 6 = ... 

0 1 2 3 4 5 6 7 8 9 10 11 12 13 14 15 16 17 18

Start at 8, add 6

8 + 6 = ... 

0 1 2 3 4 5 6 7 8 9 10 11 12 13 14 15 16 17 18

Start at 9, add 6

9 + 6 = ... 

Draw in jumps forward

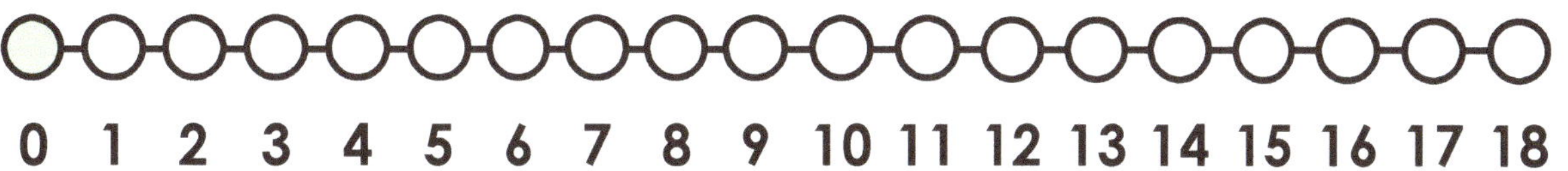

0 1 2 3 4 5 6 7 8 9 10 11 12 13 14 15 16 17 18

Start at 0, add 7

0   +   7   = ...

0 1 2 3 4 5 6 7 8 9 10 11 12 13 14 15 16 17 18

Start at 1, add 7

1   +   7   = ...

0 1 2 3 4 5 6 7 8 9 10 11 12 13 14 15 16 17 18

Start at 2, add 7

2   +   7   = ...

Draw in jumps forward

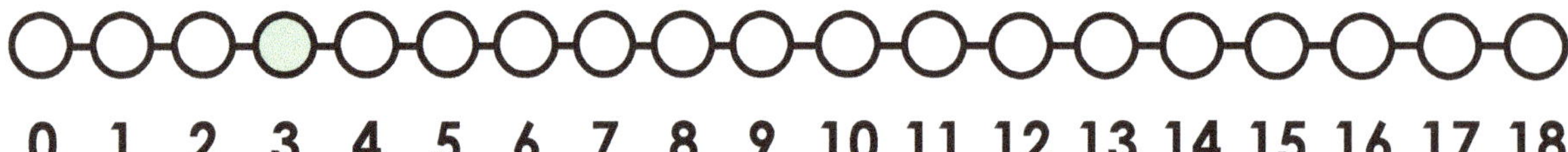

Start at 3, add 7

$$3 \quad + \quad 7 \quad = \quad \text{................}$$

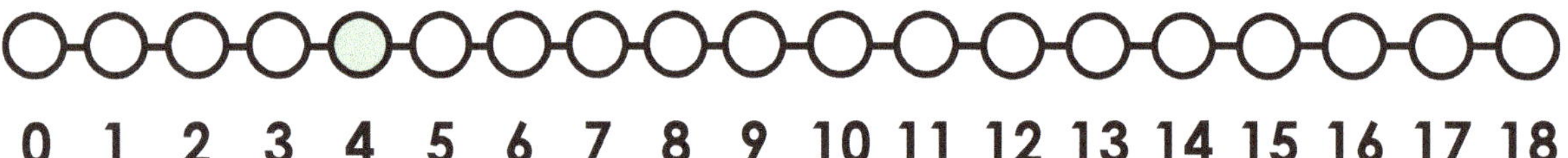

Start at 4, add 7

$$4 \quad + \quad 7 \quad = \quad \text{................}$$

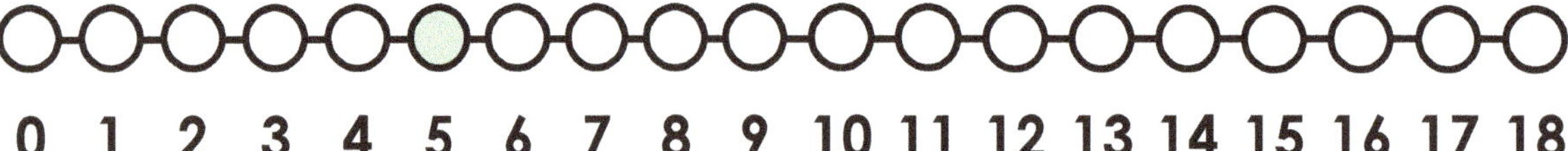

Start at 5, add 7

$$5 \quad + \quad 7 \quad = \quad \text{................}$$

Draw in jumps forward

0 1 2 3 4 5 6 7 8 9 10 11 12 13 14 15 16 17 18

Start at 6, add 7

6   +   7   =    ....

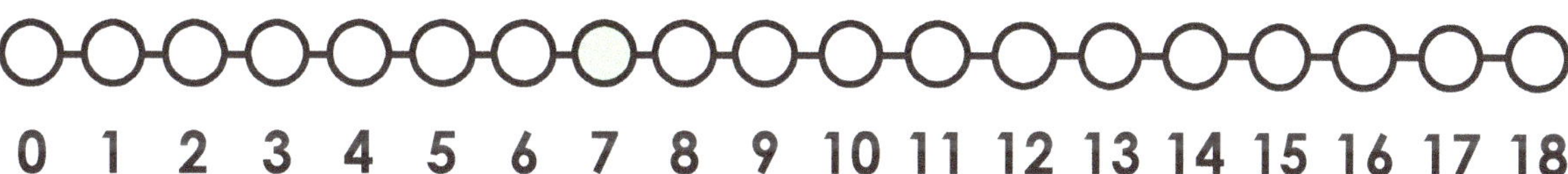

0 1 2 3 4 5 6 7 8 9 10 11 12 13 14 15 16 17 18

Start at 7, add 7

7   +   7   =    ....

---

0 1 2 3 4 5 6 7 8 9 10 11 12 13 14 15 16 17 18

Start at 8, add 7

8   +   7   = ...

Draw in jumps forward

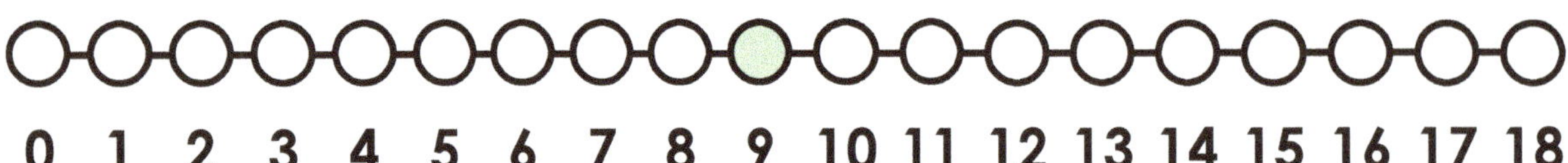

0 1 2 3 4 5 6 7 8 9 10 11 12 13 14 15 16 17 18

Start at 9, add 7

9   +   7   = ... 

0 1 2 3 4 5 6 7 8 9 10 11 12 13 14 15 16 17 18

Start at 0, add 8

0   +   8   = ... 

0 1 2 3 4 5 6 7 8 9 10 11 12 13 14 15 16 17 18

Start at 1, add 8

1   +   8   = ... 

Draw in jumps forward

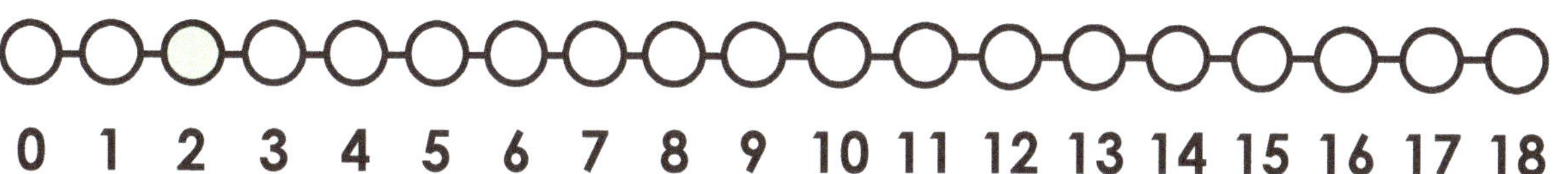

Start at 2, add 8

$$2 \quad + \quad 8 \quad = \quad \underline{\qquad}$$

Start at 3, add 8

$$3 \quad + \quad 8 \quad = \quad \underline{\qquad}$$

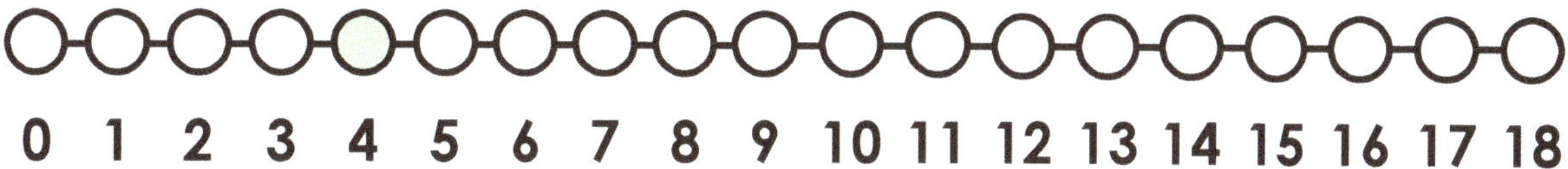

Start at 4, add 8

$$4 \quad + \quad 8 \quad = \quad \underline{\qquad}$$

Draw in jumps forward

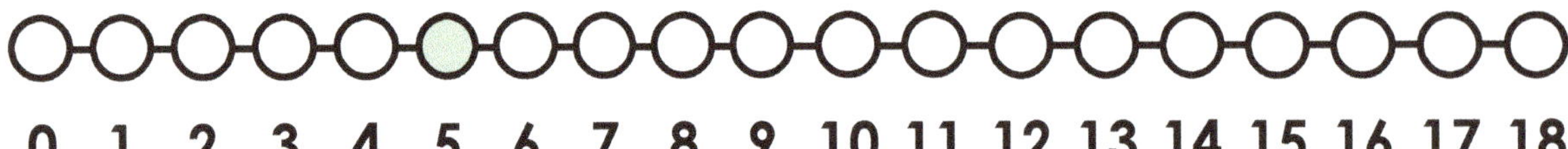

0 1 2 3 4 5 6 7 8 9 10 11 12 13 14 15 16 17 18

Start at 5, add 8

5 + 8 = ...

0 1 2 3 4 5 6 7 8 9 10 11 12 13 14 15 16 17 18

Start at 6, add 8

6 + 8 = ...

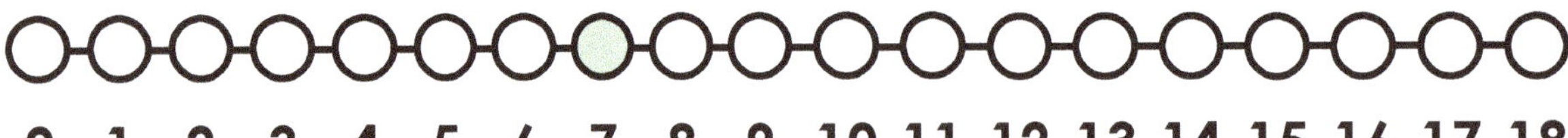

0 1 2 3 4 5 6 7 8 9 10 11 12 13 14 15 16 17 18

Start at 7, add 8

7 + 8 = ...

Draw in jumps forward

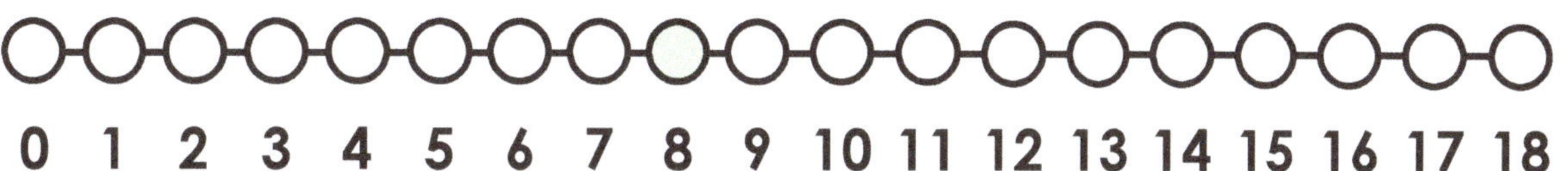

Start at 8, add 8

8 + 8 = ...

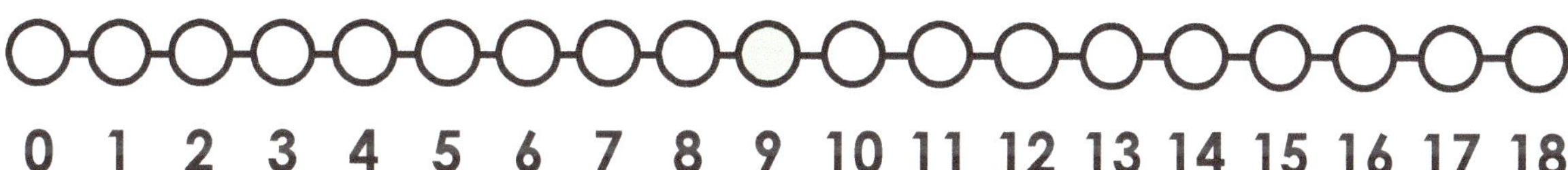

Start at 9, add 8

9 + 8 = ...

Start at 0, add 9

0 + 9 = ...

Draw in jumps forward

4

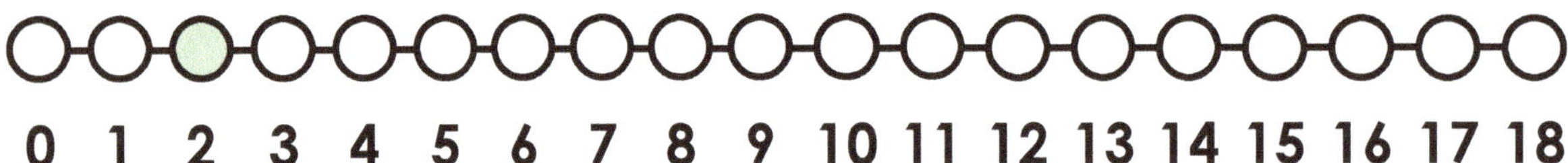

0 1 2 3 4 5 6 7 8 9 10 11 12 13 14 15 16 17 18

Start at 2, add 9

2   +   9   = ...

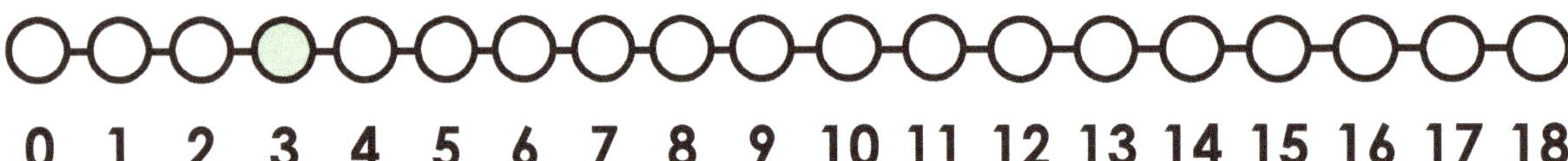

0 1 2 3 4 5 6 7 8 9 10 11 12 13 14 15 16 17 18

Start at 3, add 9

3   +   9   = ...

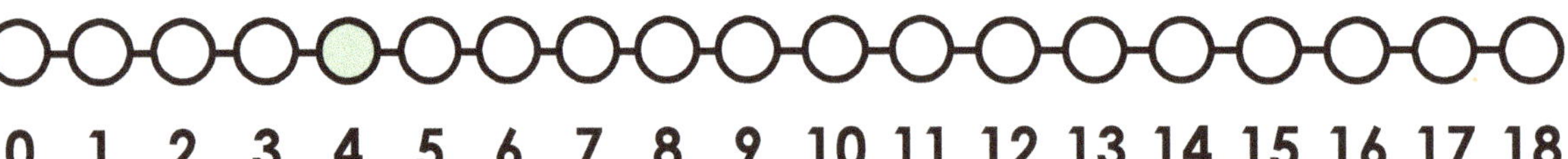

0 1 2 3 4 5 6 7 8 9 10 11 12 13 14 15 16 17 18

Start at 4, add 9

4   +   9   = ...

Draw in jumps forward

Start at 5, add 9

    5   +   9   = ...

Start at 6, add 9

    6   +   9   = ...

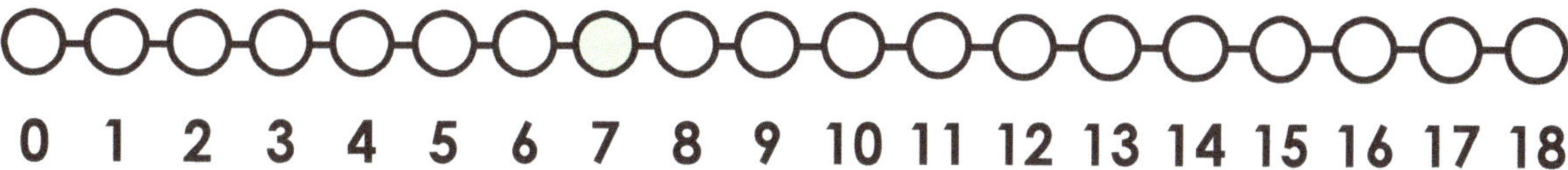

Start at 7, add 9

    7   +   9   = ...

Draw in jumps forward

4

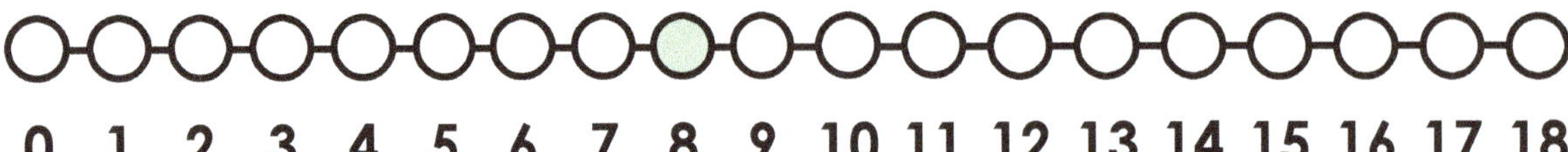

0 1 2 3 4 5 6 7 8 9 10 11 12 13 14 15 16 17 18

Start at 8, add 9

8 + 9 = ...

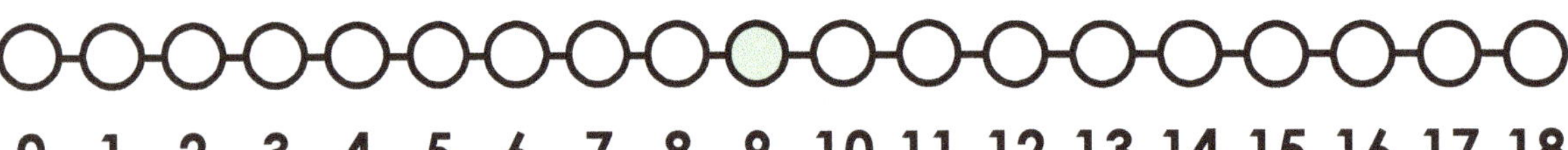

0 1 2 3 4 5 6 7 8 9 10 11 12 13 14 15 16 17 18

Start at 9, add 9

9 + 9 = ...

Problem solving

Count forward

3   +   1   = ... 

2   +   4   = ...

1   +   5   = ...

9   +   2   = ... 

12   +   1   = ...

16   +   5   = ...

10   +   10   = ... 

Problem solving

How many?

a + a = $\underline{2a}$

 +  = 2 

Count forward

$2a$ + $2a$ = ... 

$3a$ + $5a$ = ...

$2x$ + $3x$ = ...

Problem solving

Money values

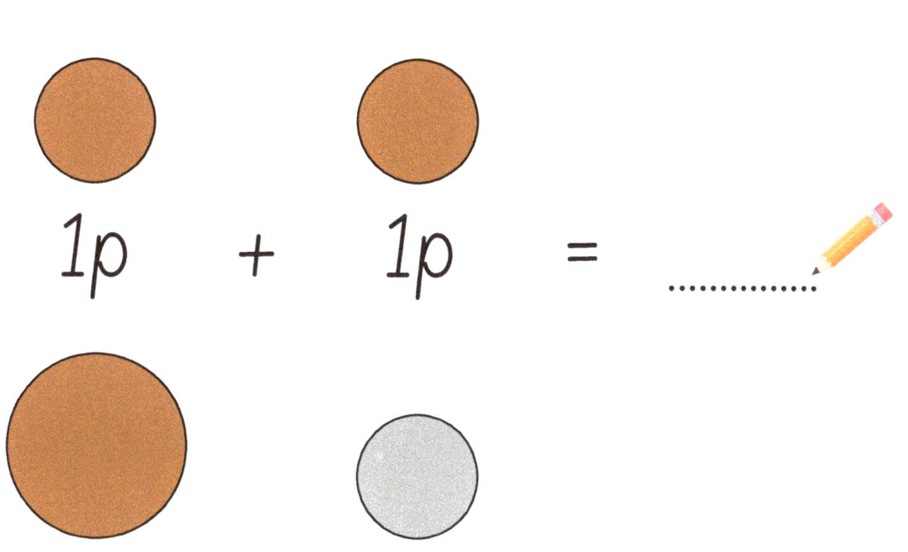

1p   +   1p   = ...

2p   +   5p   = ...

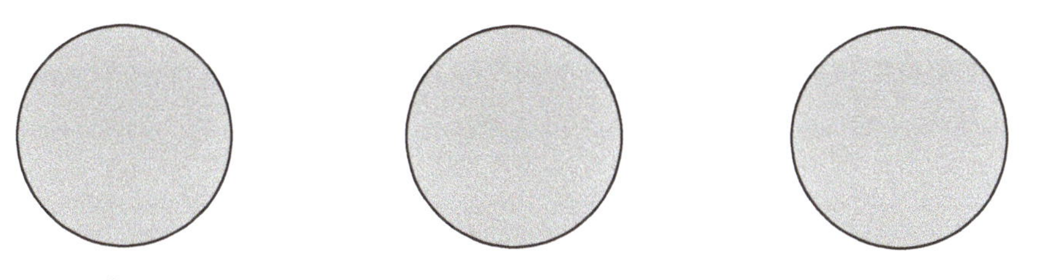

10p   +   10p   +   10p   = ...

20p   +   10p   +   5p   = ...

Notes

86

Notes

87

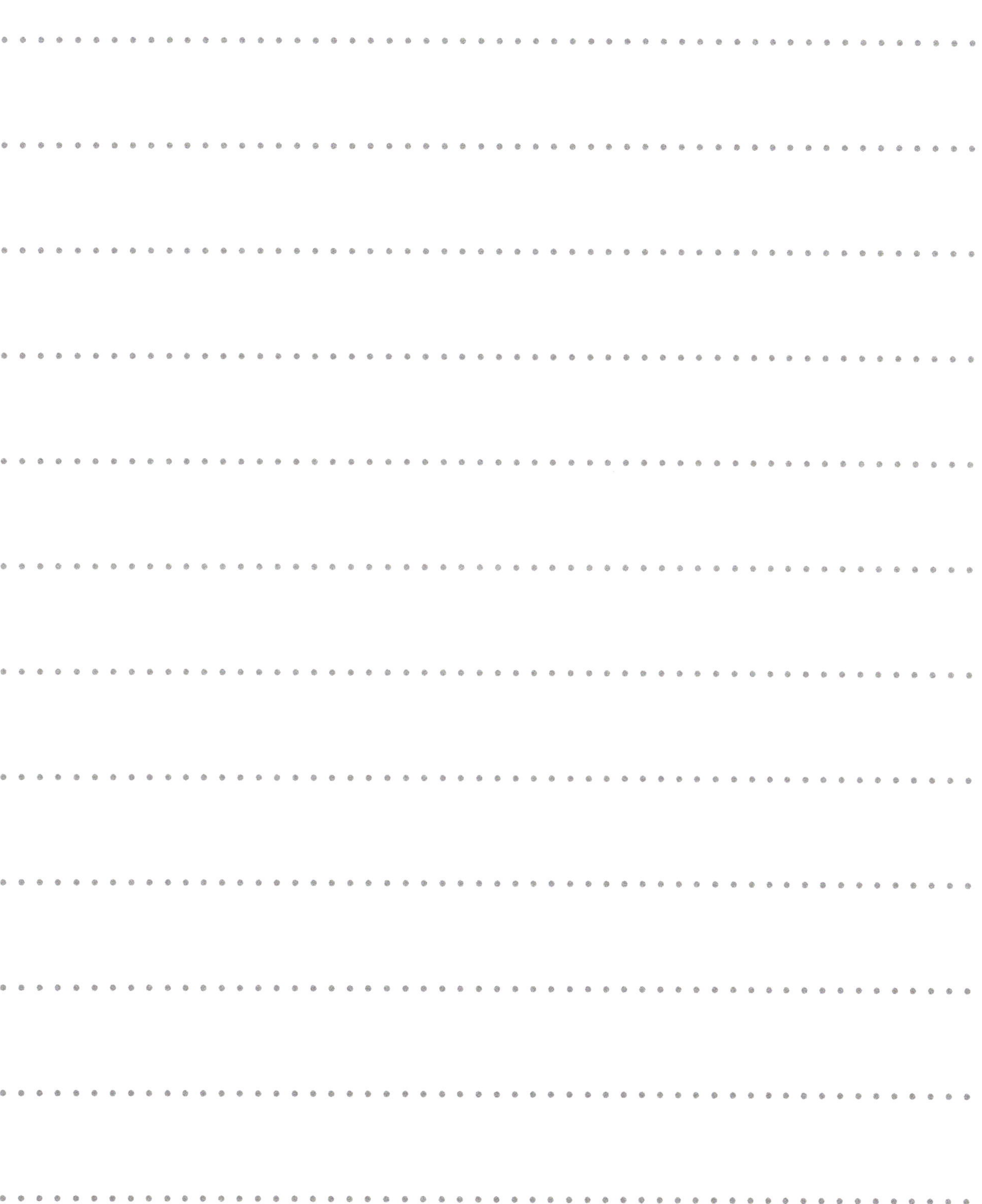

Unit 5

Take away – Number line

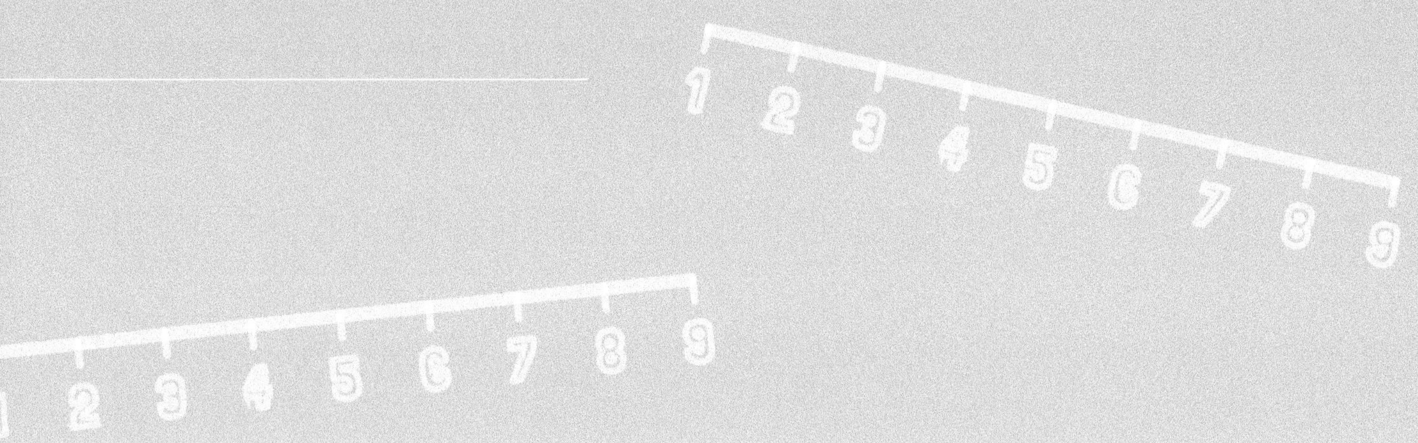

Teacher Note

Number line

Reinforcing number bonds to 10

Outcome: Number, money and measurement

Strand: Add and subtract

Level: A – add and subtract – mentally for numbers 0 to 10.
B – add and subtract – mentally for numbers 0 to 20.

Aims

The pupils should be able to subtract for numbers up to 10.

Introduction

The pupils should be introduced to the number line. Number lines could be made from card and one given to each pupil. Pupils should be instructed to add in the 'jumps' backward on the number line. They should be encouraged to count out loud and then complete the sum.

Draw in jumps backwards

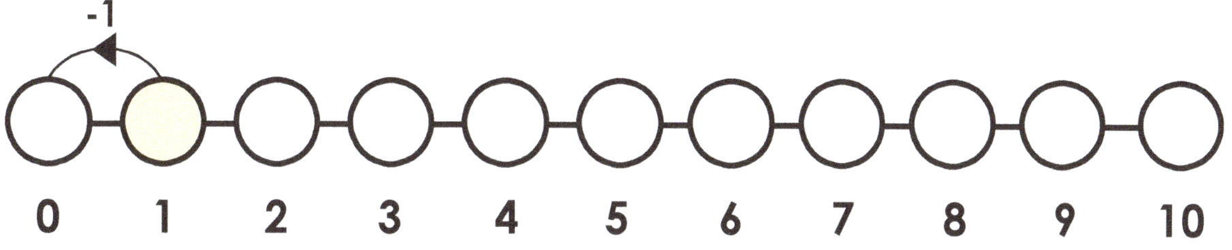

Start at 1, take away 1

$$1 - 1 = \text{................}$$

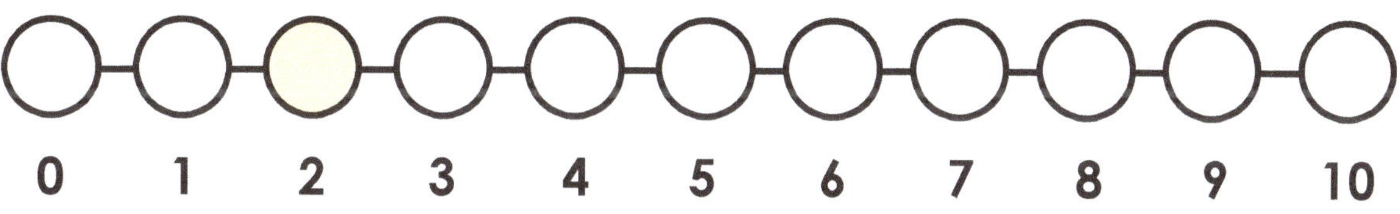

Start at 2, take away 1

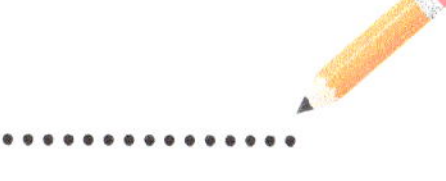

$$2 - 1 = \text{................}$$

Start at 3, take away 1

$$3 - 1 = \text{................}$$

Draw in jumps backwards

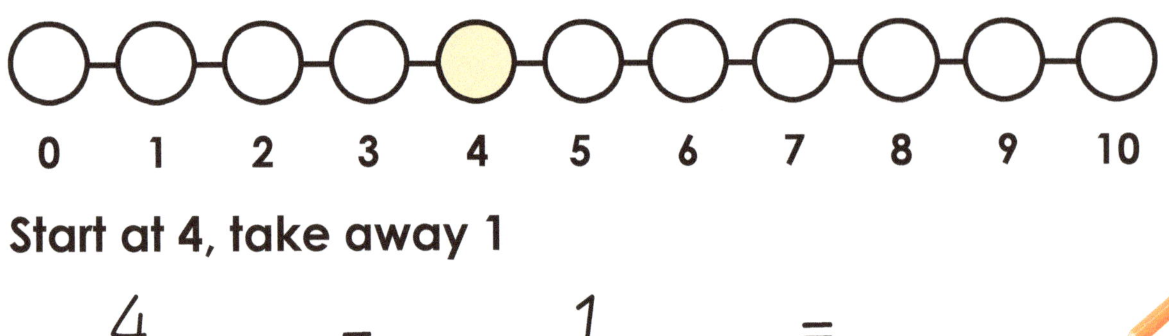

Start at 4, take away 1

4   –   1   = ...

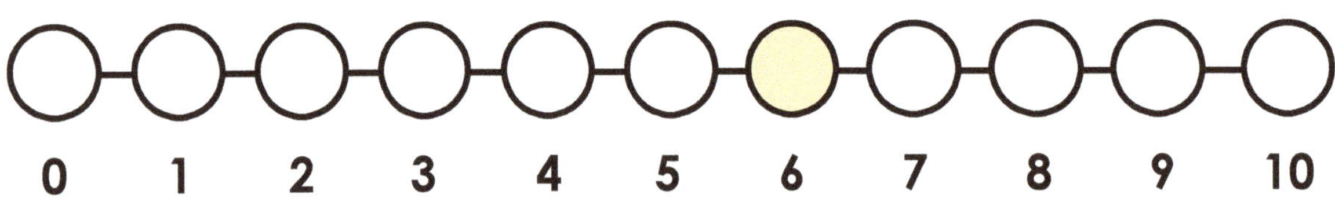

Start at 5, take away 1

5   –   1   = ...

Start at 6, take away 1

6   –   1   = ...

Draw in jumps backwards

0 1 2 3 4 5 6 7 8 9 10

Start at 7, take away 1

7 – 1 = ...

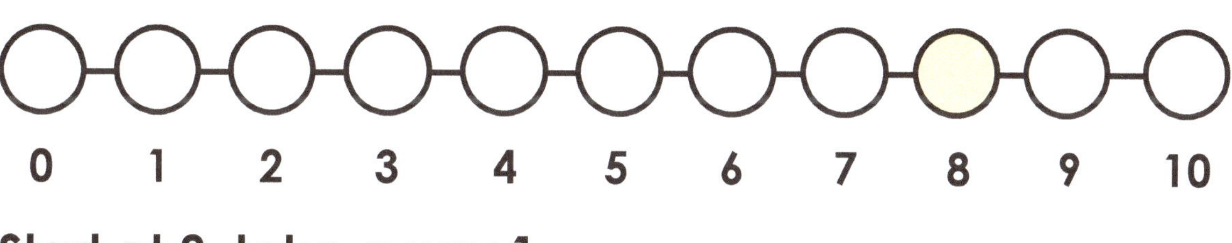

0 1 2 3 4 5 6 7 8 9 10

Start at 8, take away 1

8 – 1 = ...

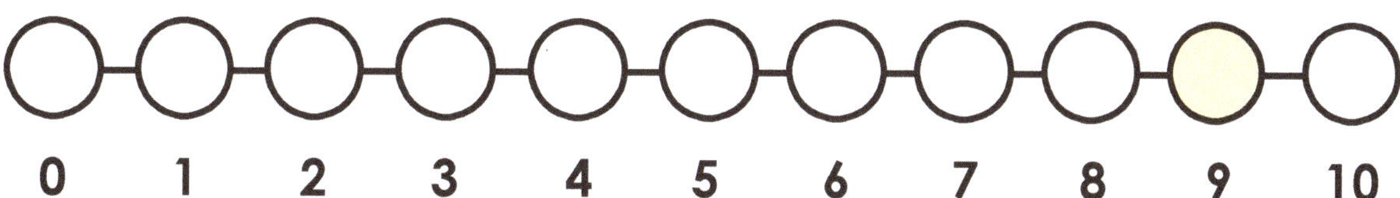

0 1 2 3 4 5 6 7 8 9 10

Start at 9, take away 1

9 – 1 = ...

Draw in jumps backwards

Start at 2, take away 2

2   –   2   = ...

Start at 3, take away 2

3   –   2   = ...

Start at 4, take away 2

4   –   2   = ...

Draw in jumps backwards

0 1 2 3 4 5 6 7 8 9 10 11

Start at 5, take away 2

$$5 \quad - \quad 2 \quad = \quad \underline{\quad\quad}$$

0 1 2 3 4 5 6 7 8 9 10 11

Start at 6, take away 2

$$6 \quad - \quad 2 \quad = \quad \underline{\quad\quad}$$

0 1 2 3 4 5 6 7 8 9 10 11

Start at 7, take away 2

$$7 \quad - \quad 2 \quad = \quad \underline{\quad\quad}$$

Draw in jumps backwards

Start at 8, take away 2

8 – 2 = ...

Start at 9, take away 2

9 – 2 = ...

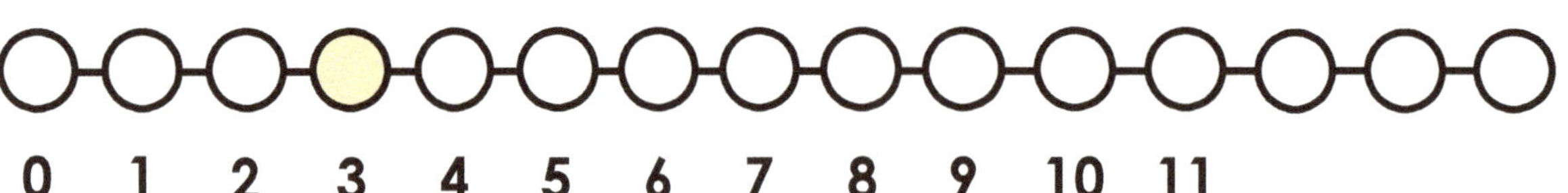

Start at 3, take away 3

3 – 3 = ...

Draw in jumps backwards

Start at 4, take away 3

4   –   3   = ...

Start at 5, take away 3

5   –   3   = ...

Start at 6, take away 3

6   –   3   = ...

Draw in jumps backwards

0 1 2 3 4 5 6 7 8 9 10 11 12

Start at 7, take away 3

7 − 3 = ...

0 1 2 3 4 5 6 7 8 9 10 11 12

Start at 8, take away 3

8 − 3 = ...

0 1 2 3 4 5 6 7 8 9 10 11 12

Start at 9, take away 3

9 − 3 = ...

Draw in jumps backwards

Start at 4, take away 4

$$4 \quad - \quad 4 \quad = \quad \underline{\hspace{2cm}}$$

Start at 5, take away 4

$$5 \quad - \quad 4 \quad = \quad \underline{\hspace{2cm}}$$

Start at 6, take away 4

$$6 \quad - \quad 4 \quad = \quad \underline{\hspace{2cm}}$$

Draw in jumps backwards

Start at 7, take away 4

$$7 \quad - \quad 4 \quad = \quad \text{...............}$$

Start at 8, take away 4

$$8 \quad - \quad 4 \quad = \quad \text{...............}$$

Start at 9, take away 4

$$9 \quad - \quad 4 \quad = \quad \text{...............}$$

Draw in jumps backwards

Start at 5, take away 5

5   –   5   = ...

Start at 6, take away 5

6   –   5   = ...

Start at 7, take away 5

7   –   5   = ...

Draw in jumps backwards

Start at 8, take away 5

8   –   5   = ...

Start at 9, take away 5

9   –   5   = ...

Start at 6, take away 6

6   –   6   = ...

Draw in jumps backwards

0 1 2 3 4 5 6 7 8 9 10 11 12 13

Start at 7, take away 6

7 – 6 = 

0 1 2 3 4 5 6 7 8 9 10 11

Start at 8, take away 6

8 – 6 = 

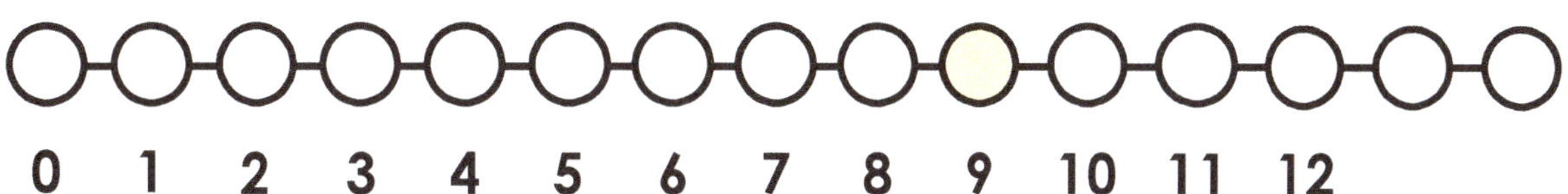

0 1 2 3 4 5 6 7 8 9 10 11 12

Start at 9, take away 6

9 – 6 =

Draw in jumps backwards

0 1 2 3 4 5 6 7 8 9 10 11 12 13

Start at 7, take away 7

7 – 7 = ... 

0 1 2 3 4 5 6 7 8 9 10 11 12 13 14

Start at 8, take away 7

8 – 7 = ... 

0 1 2 3 4 5 6 7 8 9 10 11 12 13 14

Start at 9, take away 7

9 – 7 = ... 

Draw in jumps backwards

Start at 8, take away 6

$$8 \quad - \quad 8 \quad = \quad \underline{\hspace{1.5cm}}$$

Start at 9, take away 8

$$9 \quad - \quad 8 \quad = \quad \underline{\hspace{1.5cm}}$$

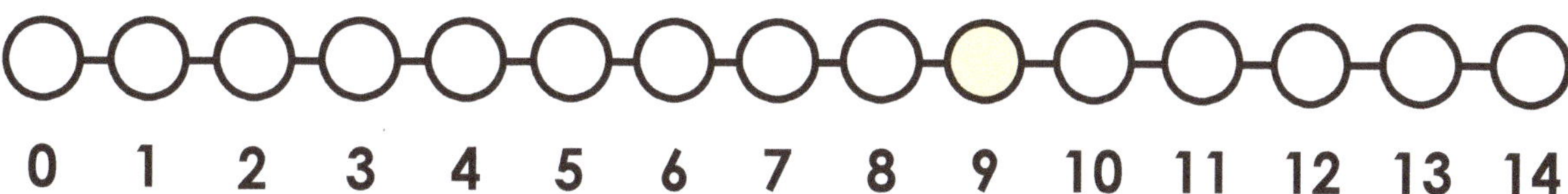

Start at 9, take away 9

$$9 \quad - \quad 9 \quad = \quad \underline{\hspace{1.5cm}}$$

Notes

Notes

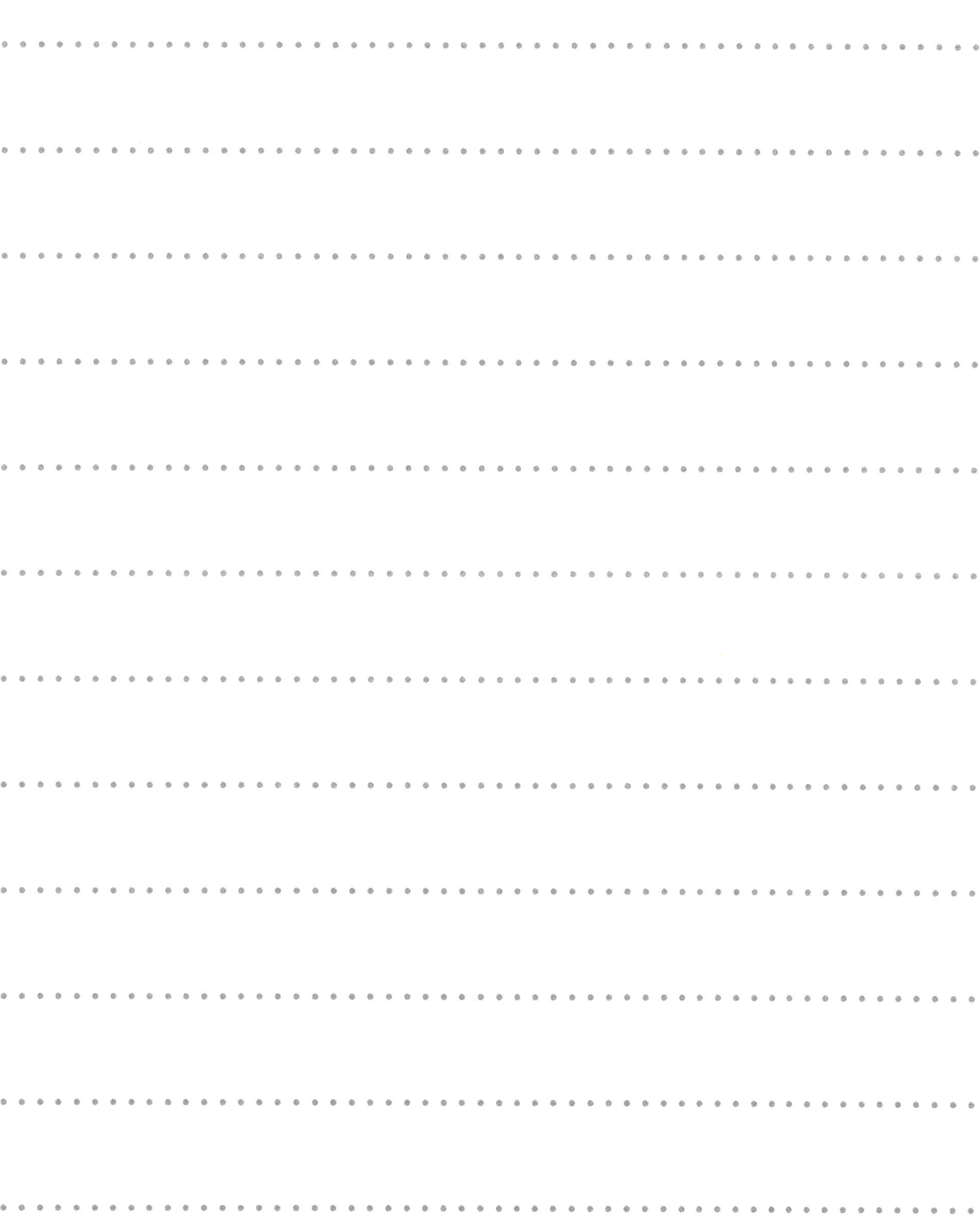

5

Unit 6

Multiply and Divide

Teacher Note

Multiple and divide

Reinforcing number bonds to 10

Outcome: Number, money and measurement

Strand: Multiple and divide

Level: A – multiply and divide mentally by 2, 3, 4, 5 and 10.
B – multiply and divide mentally within the confines of all tables to 10.

Aims

The pupils should know multiples of 2, 3, 10, 3 and 4 at level B. (6, 7, 8, and 9 at level C)

Introduction to 2x table

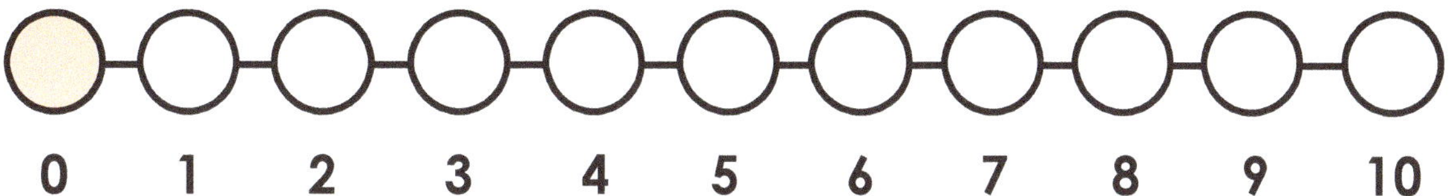

0   1   2   3   4   5   6   7   8   9   10

Start at 0, colour every 2nd step.

Complete the sequence:

0,   2,   4, ..., ..., ...

Colour the even numbers									
1	2	3	4	5	6	7	8	9	10
11	12	13	14	15	16	17	18	19	20
21	22	23	24	25	26	27	28	29	30
31	32	33	34	35	36	37	38	39	40
41	42	43	44	45	46	47	48	49	50
51	52	53	54	55	56	57	58	59	60
61	62	63	64	65	66	67	68	69	70
71	72	73	74	75	76	77	78	79	80
81	82	83	84	85	86	87	88	89	90
91	92	93	94	95	96	97	98	99	100

Write the numbers in order

2

4

18

12

14

0

8

6

16

10

20

x2 Join the dots then colour the picture

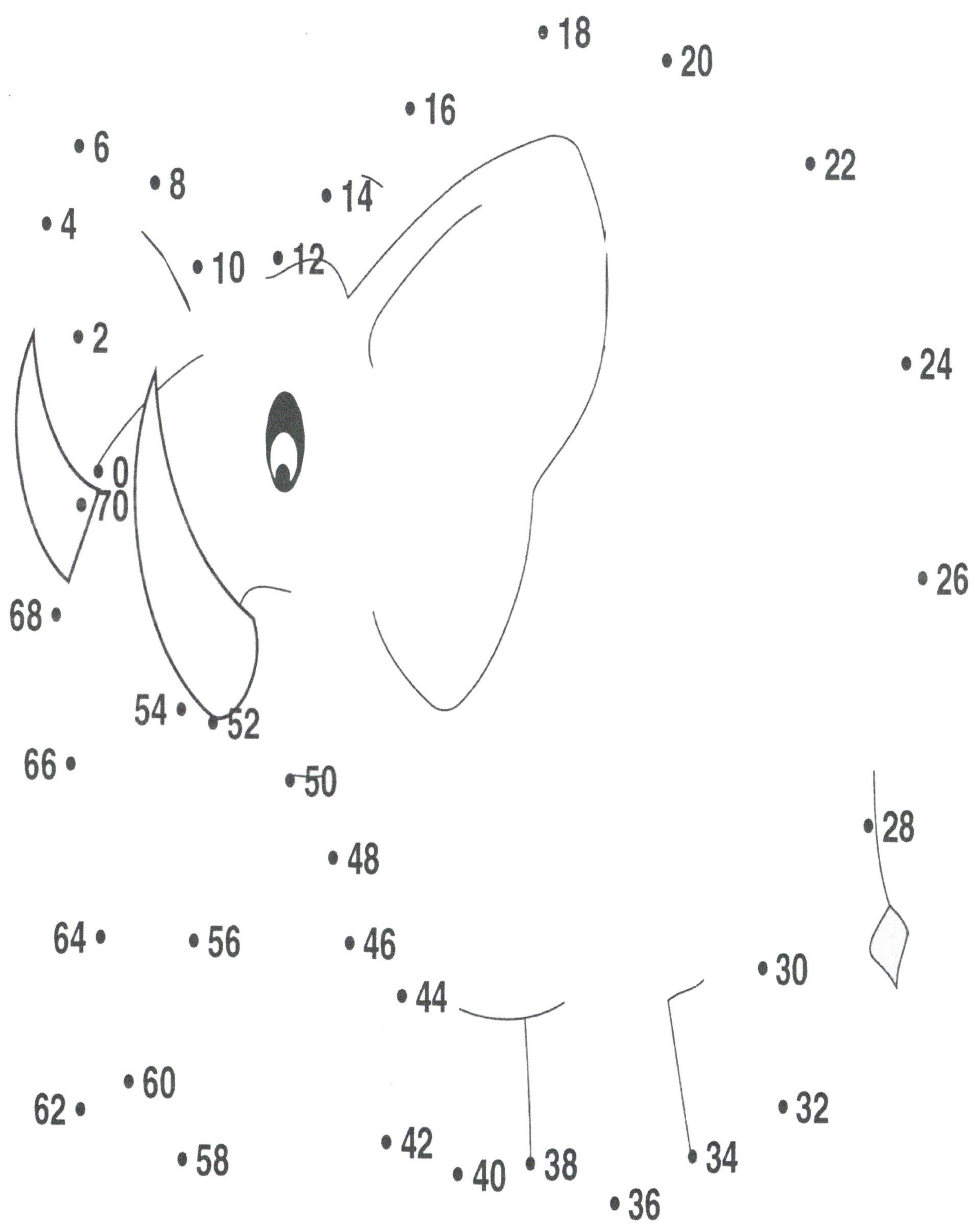

Introduction to 3x table

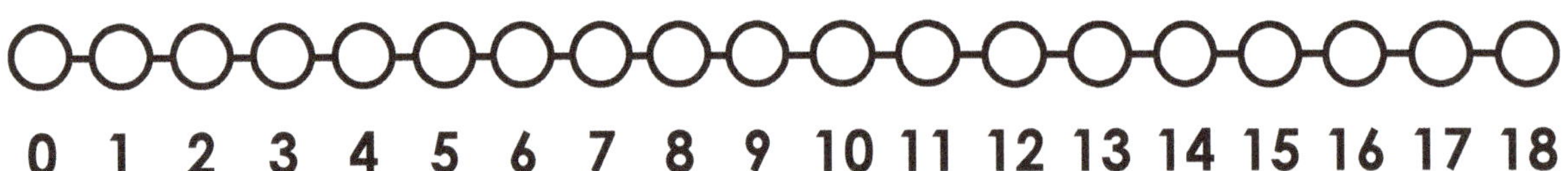

0 1 2 3 4 5 6 7 8 9 10 11 12 13 14 15 16 17 18

6

Start at 0, colour every 3nd step.
Complete the sequence:

0,　　3,　　6,　　....,　　....,　　....,　　....,　　....,　　....,　　....,　　30.

Write the numbers in order:

6

12

24

30

3

9

21

0

15

18

27

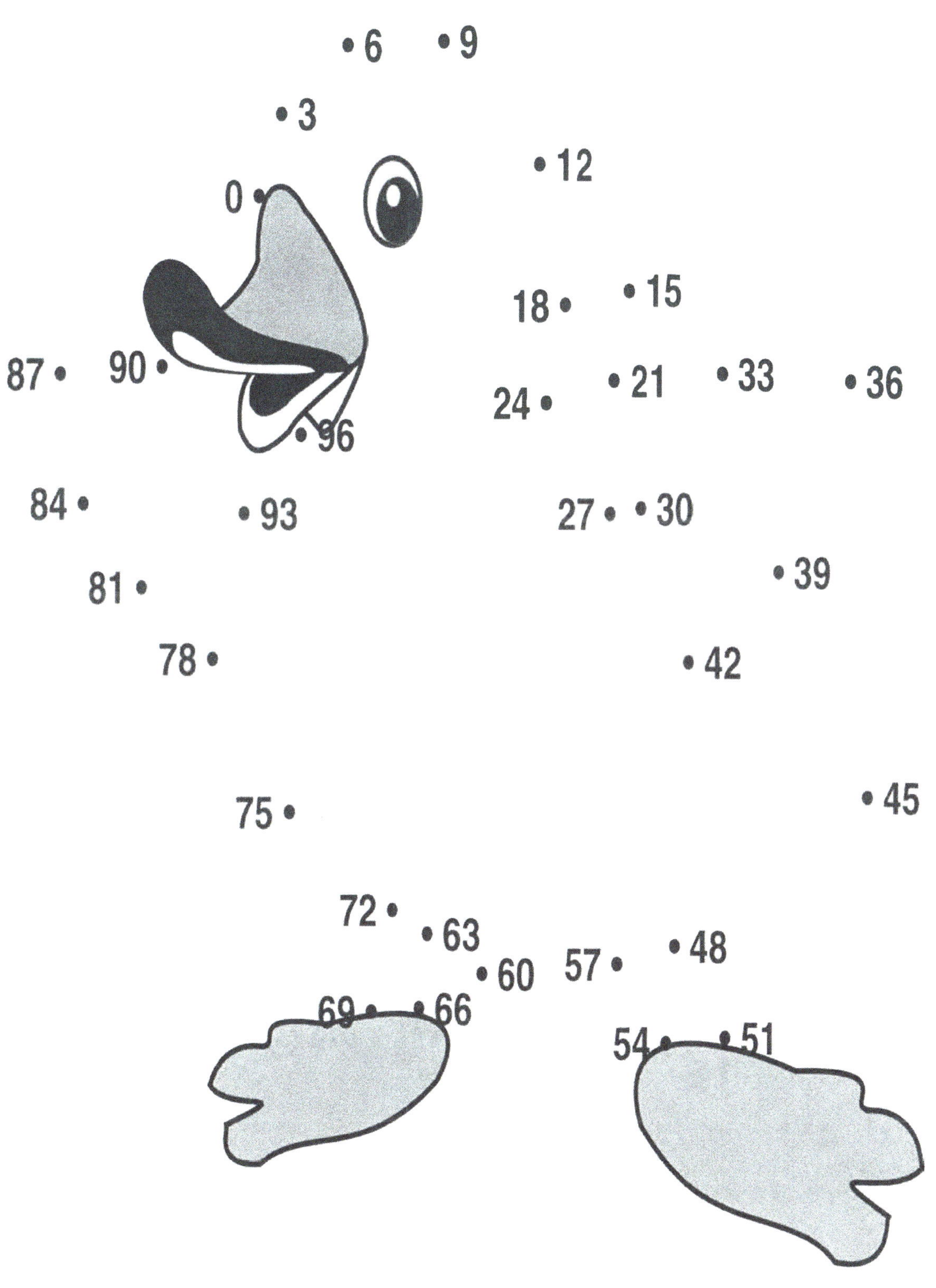

Find the number divisible by 3

Find the numbers divisible by 3 then colour the square (hint: start at 3 then go up in steps of 3)

1	2	3	4	5	6	7	8	9	10
11	12	13	14	15	16	17	18	19	20
21	22	23	24	25	26	27	28	29	30
31	32	33	34	35	36	37	38	39	40
41	42	43	44	45	46	47	48	49	50
51	52	53	54	55	56	57	58	59	60
61	62	63	64	65	66	67	68	69	70
71	72	73	74	75	76	77	78	79	80
81	82	83	84	85	86	87	88	89	90
91	92	93	94	95	96	97	98	99	100

Multiply

How Many?

 **+**  **+**  **+**  **+** 

3 sets of 1 

3 x 1 

 **+**  **+**  **+**  **+** 

3 sets of 2 

3 x 2

Introduction to 4x table

0 1 2 3 4 5 6 7 8 9 10 11 12 13 14 15 16 17 18 19 20

6

Start at 0, colour every 4th step.
Complete the sequence:

0,   4,   8, ..., ..., ..., ..., ..., ..., ...,   40.

Write the numbers in order:

0

8

36

4

12

24

16

32

28

20

40

Find the numbers divisible by 4

Find the numbers divisible by 4 then colour the square
(hint: start at 3 then go up in steps of 4)

1	2	3	4	5	6	7	8	9	10
11	12	13	14	15	16	17	18	19	20
21	22	23	24	25	26	27	28	29	30
31	32	33	34	35	36	37	38	39	40
41	42	43	44	45	46	47	48	49	50
51	52	53	54	55	56	57	58	59	60
61	62	63	64	65	66	67	68	69	70
71	72	73	74	75	76	77	78	79	80
81	82	83	84	85	86	87	88	89	90
91	92	93	94	95	96	97	98	99	100

Level B Introduction to 5x table

0 1 2 3 4 5 6 7 8 9 10 11 12 13 14 15 16 17 18 19 20

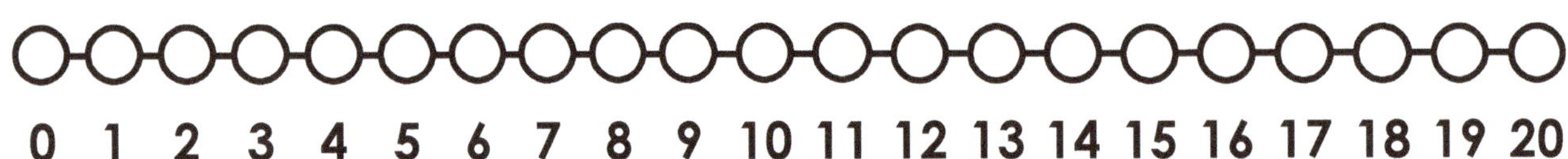

Start at 0, colour every 5th step.

Complete the sequence:

0, 5, 10, ..., ..., ..., ..., ..., ..., ..., 50.

Write the numbers in order:

5 20

35

15 50

0

25

10

45 30

40

Find the numbers divisible by 5

Find the numbers divisible by 5 then colour the square (hint: start at 3 then go up in steps of 5)

1	2	3	4	5	6	7	8	9	10
11	12	13	14	15	16	17	18	19	20
21	22	23	24	25	26	27	28	29	30
31	32	33	34	35	36	37	38	39	40
41	42	43	44	45	46	47	48	49	50
51	52	53	54	55	56	57	58	59	60
61	62	63	64	65	66	67	68	69	70
71	72	73	74	75	76	77	78	79	80
81	82	83	84	85	86	87	88	89	90
91	92	93	94	95	96	97	98	99	100

Level B Introduction to 10x table

Complete the sequence:

0,   10,   20, ..., ..., ..., ..., ..., ..., ..., 100.

Write the numbers in order:

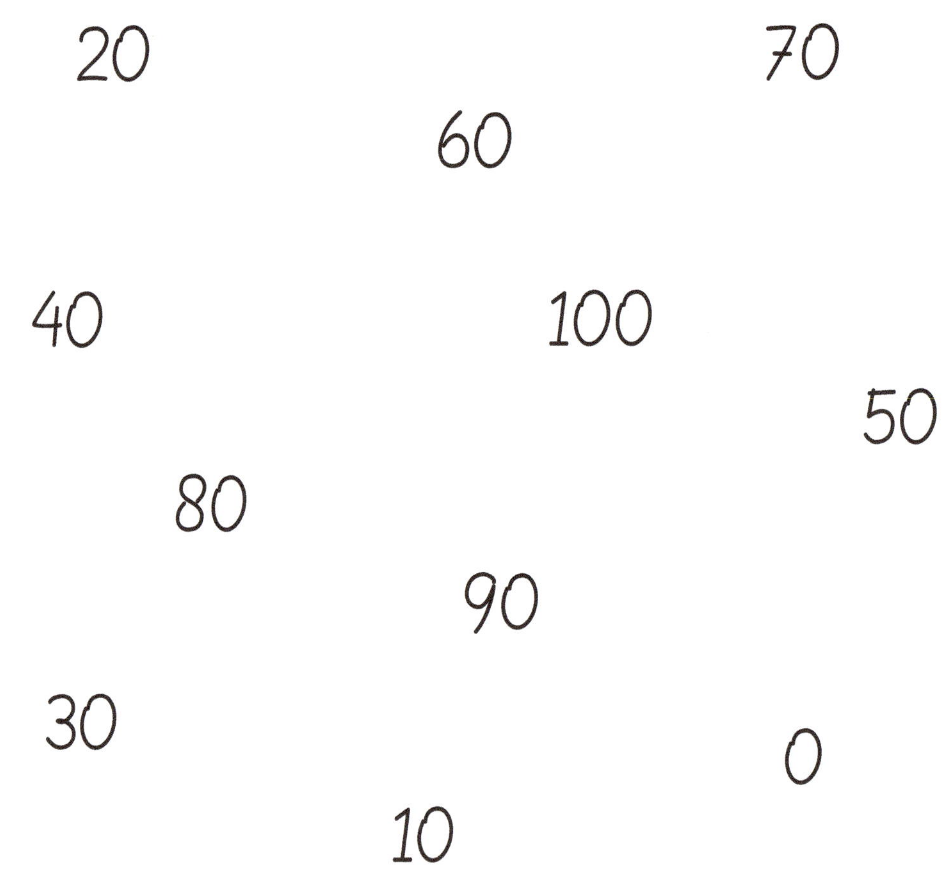

Find the numbers divisible by 10

Find the numbers divisible by 10 then colour the square
(hint: start at 3 then go up in steps of 10)

1	2	3	4	5	6	7	8	9	10
11	12	13	14	15	16	17	18	19	20
21	22	23	24	25	26	27	28	29	30
31	32	33	34	35	36	37	38	39	40
41	42	43	44	45	46	47	48	49	50
51	52	53	54	55	56	57	58	59	60
61	62	63	64	65	66	67	68	69	70
71	72	73	74	75	76	77	78	79	80
81	82	83	84	85	86	87	88	89	90
91	92	93	94	95	96	97	98	99	100

Level C Introduction to 6x table

Complete the sequence:

0,   6,   12, ..., ..., ..., ..., ..., ..., ...,   60.

6

Write the numbers in order:

30

54

36

6

48

0

60

12

42

24

18

Find the numbers divisible by 6

Find the numbers divisible by 6 then colour the square
(hint: start at 3 then go up in steps of 6)

6

1	2	3	4	5	6	7	8	9	10
11	12	13	14	15	16	17	18	19	20
21	22	23	24	25	26	27	28	29	30
31	32	33	34	35	36	37	38	39	40
41	42	43	44	45	46	47	48	49	50
51	52	53	54	55	56	57	58	59	60
61	62	63	64	65	66	67	68	69	70
71	72	73	74	75	76	77	78	79	80
81	82	83	84	85	86	87	88	89	90
91	92	93	94	95	96	97	98	99	100

Level C Introduction to 7x table

Complete the sequence:

0,　　7,　　14,　　.....,　　.....,　　.....,　　.....,　　.....,　　.....,　　.....,　　70.

Write the numbers in order:

Find the numbers divisible by 7

Find the numbers divisible by 7 then colour the square
(hint: start at 3 then go up in steps of 7)

1	2	3	4	5	6	7	8	9	10
11	12	13	14	15	16	17	18	19	20
21	22	23	24	25	26	27	28	29	30
31	32	33	34	35	36	37	38	39	40
41	42	43	44	45	46	47	48	49	50
51	52	53	54	55	56	57	58	59	60
61	62	63	64	65	66	67	68	69	70
71	72	73	74	75	76	77	78	79	80
81	82	83	84	85	86	87	88	89	90
91	92	93	94	95	96	97	98	99	100

Level C Introduction to 8x table

Complete the sequence:

0,   8,   16, ..., ..., ..., ..., ..., ..., ...,   80.

6 Write the numbers in order:

80

56

16

48

32

40

0

24

72

64

8

Find the numbers divisible by 8

Find the numbers divisible by 8 then colour the square
(hint: start at 3 then go up in steps of 8)

1	2	3	4	5	6	7	8	9	10
11	12	13	14	15	16	17	18	19	20
21	22	23	24	25	26	27	28	29	30
31	32	33	34	35	36	37	38	39	40
41	42	43	44	45	46	47	48	49	50
51	52	53	54	55	56	57	58	59	60
61	62	63	64	65	66	67	68	69	70
71	72	73	74	75	76	77	78	79	80
81	82	83	84	85	86	87	88	89	90
91	92	93	94	95	96	97	98	99	100

Level C Introduction to 9x table

Complete the sequence:

0,   9,   18, ..., ..., ..., ..., ..., ..., ...,   90.

6

Write the numbers in order:

72

45

90

0

81

27

9

36

54

63

18

Find the numbers divisible by 9

Find the numbers divisible by 9 then colour the square
(hint: start at 3 then go up in steps of 9)

1	2	3	4	5	6	7	8	9	10
11	12	13	14	15	16	17	18	19	20
21	22	23	24	25	26	27	28	29	30
31	32	33	34	35	36	37	38	39	40
41	42	43	44	45	46	47	48	49	50
51	52	53	54	55	56	57	58	59	60
61	62	63	64	65	66	67	68	69	70
71	72	73	74	75	76	77	78	79	80
81	82	83	84	85	86	87	88	89	90
91	92	93	94	95	96	97	98	99	100

Notes

Notes

Unit 7

Rounding 2 digit numbers to the nearest 10

Teacher Note

Rounding 2 digit whole numbers to the nearest 10

Outcome: Number, money and measurement

Strand: Round numbers

Target: Round 2 whole digit whole numbers to the nearest 10.

Level: B.

Aims

To round two digit whole numbers to the nearest ten.

Prior knowledge

Pupils should have met 2 digit whole numbers in other contexts.

Round to the nearest 10

Example: Round 21 to the nearest 10

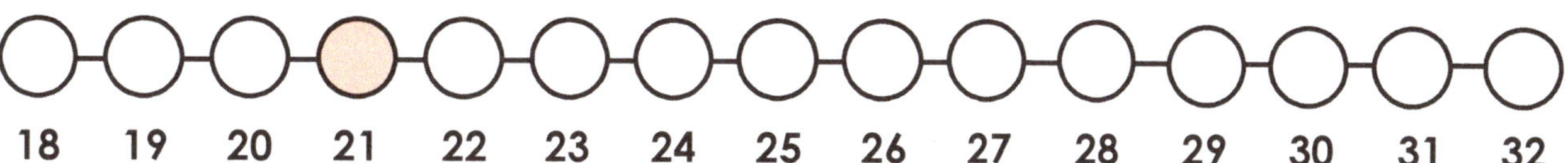

21 lies between 20 and 30 as shown on the number line.

It is nearest to 20

so   21   20   30   **21 is rounded to 20**

Now

1. Round 24 to the nearest 10.

   a. Find 24 on the number line

   b. Which two tens does 24 lie between?

   c. Circle the ten that 24 is nearest to.

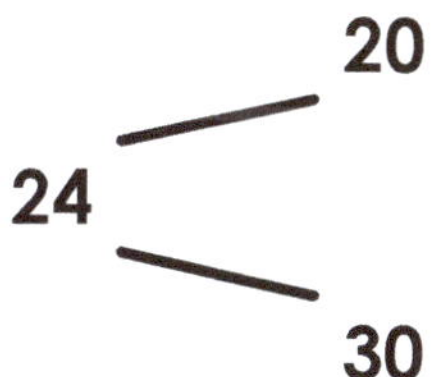

2. Round 28 to the nearest 10.

    a. Find 28 on the number line

    b. Which two tens does 28 lie between?

    c. Circle the ten that 28 is nearest to.

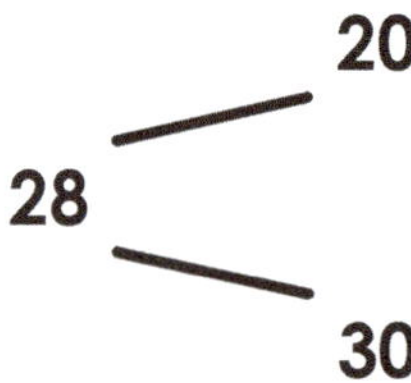

7

3. Circle the ten that each of the following is nearest to.

a. 21   20 / 30

b. 16   20 / 30

c. 11   20 / 30

d. 19   20 / 30

e. 22   20 / 30

f. 29   20 / 30

g. 31   30 / 40

h. 52   50 / 60

i. 43   40 / 50

j. 34   30 / 40

k. 38   30 / 40

l. 49   40 / 50

Round the following to the nearest 10

The first one has been done for you.

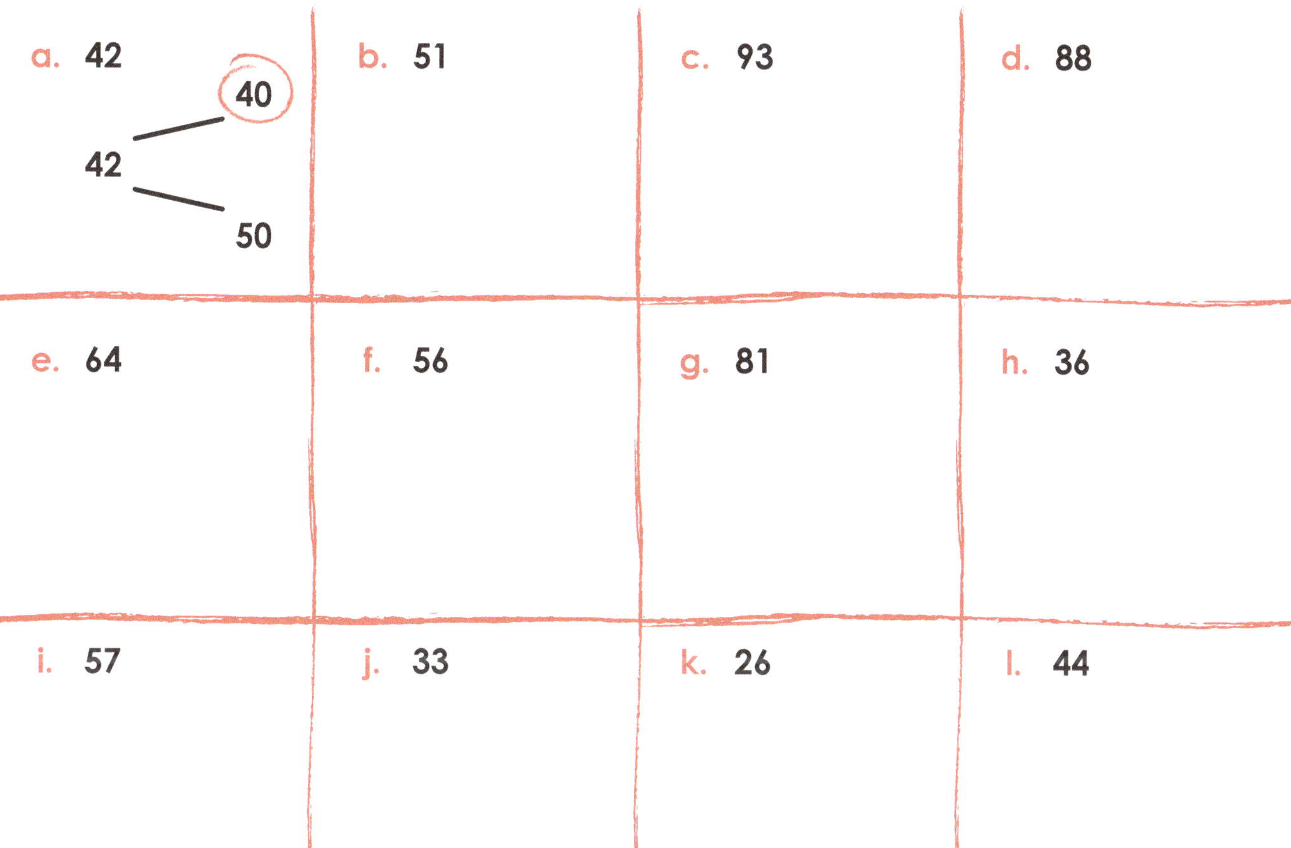

Round the following amounts to the nearest 10.

a. 49p ....   b. 31p ....   c. 56p ....   d. 8p ....

e. 93p ....   f. 42p ....   g. 57p ....   h. 68p ....

Round the following measurements to the nearest 10.

a. 22cm ....   b. 19cm ....   c. 6cm ....   d. 34cm ....

e. 42cm ....   f. 28cm ....   g. 93cm ....   h. 88cm ....

Problems

1. Mark buys two sweets.
One costs 32p, the other 39p.

a. Round each value to the nearest 10p.

b. Estimate the total cost of the sweets.

c. Find the exact value that Mark has to pay.

d. From the coins below, choose which ones Mark should pay with.

e. How much money does Mark have left?

2. A small toy costs 94p and a pencil costs 44p.

a. Round each value to the nearest 10p.

b. Estimate the total cost.

c. Find the exact value of the toy and the pencil.

d. Which works out to be more - the estimate or the exact value?

e. The boy gives the shopkeeper £1.50. How much change does he get?

3. Two new pencils measure 16cm each.

a.   Round the length of the pencils to the nearest 10cm.

b.   Estimate the total length of the two pencils.

c.   Work out the *exact* length of the two pencils.

d.   Is the estimate *more* or *less* than the exact length?

e.   Round the widths of the boxes to the nearest 10cm.

4. 2 boxes of cereal measures 26cm and 27cm wide.

a.   Estimate the total width of the two boxes.

b.   Work out the *exact* width of the boxes when put side by side.

c.   If a shelf is 60cm wide, will the boxes fit on the shelf when placed side by side.

Notes

7

Notes

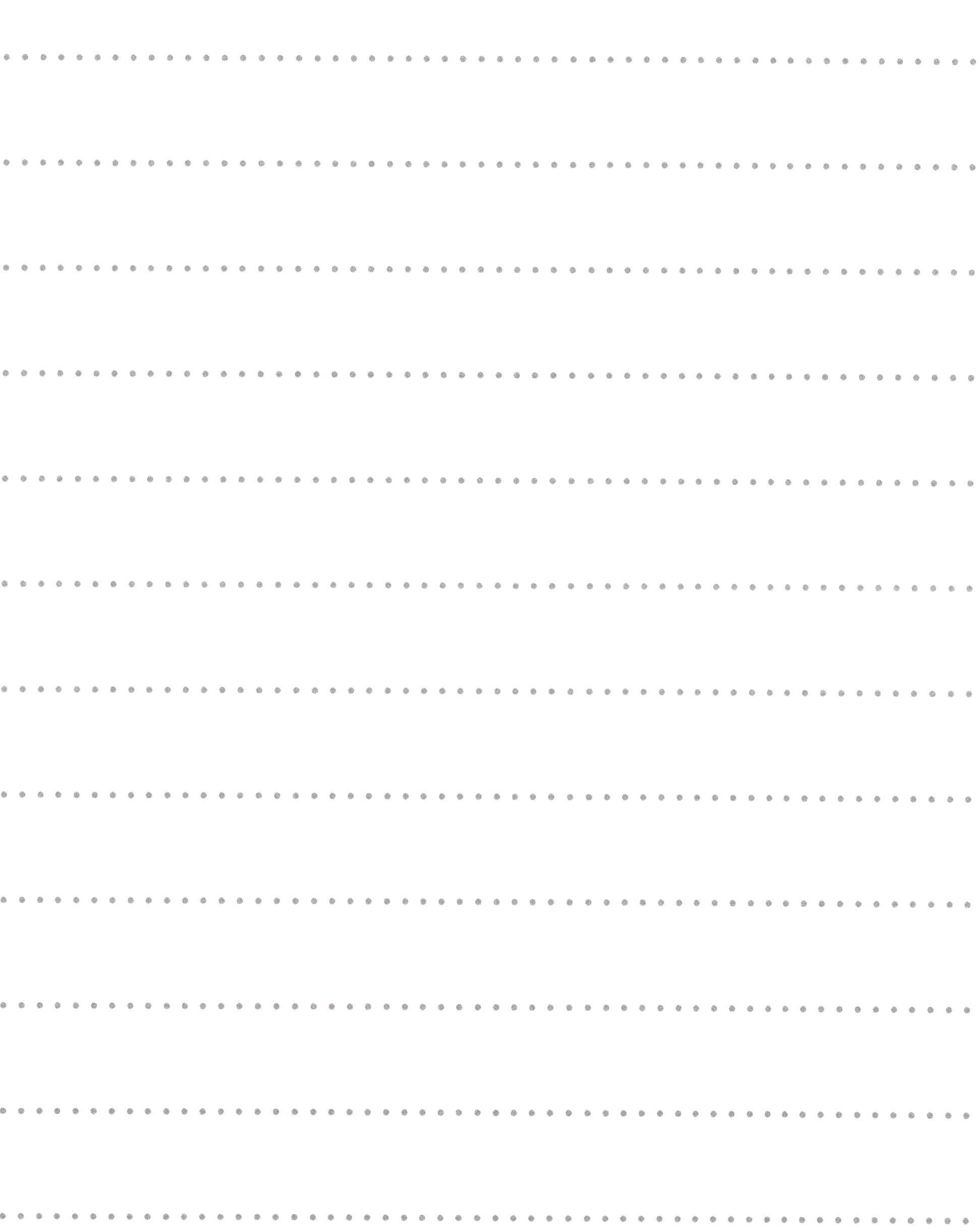

Unit 8

Rounding 3 digit numbers to the nearest 10

Teacher Note

Rounding 3 digit whole numbers to the nearest 10 (numbers, money and measurement plus problems)

Outcome: Number, money and measurement

Strand: Round numbers

Target: Round 2 whole digit whole numbers to the nearest 10.

Level: C.

Aims

To round three digit whole numbers to the nearest ten.

Prior knowledge

Pupils should have met 3 digit whole numbers in other contexts.

Round to the nearest 10

Example: Round 132 to the nearest 10

132 lies between 130 and 140 as shown on the number line.

It is nearest to 130

132 is rounded to 130

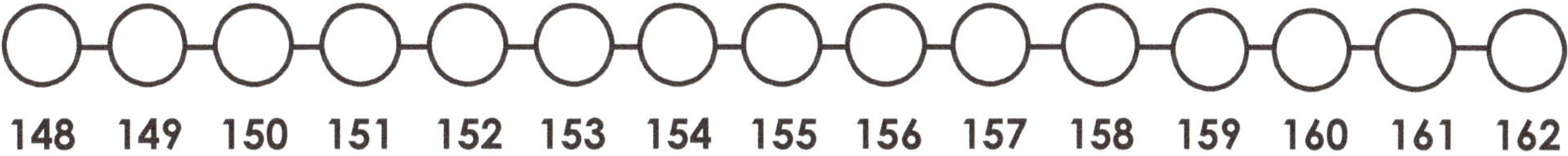

a. **Find 152 on the number line**

b. **Which two tens does 152 lie between?**

c. **Circle the ten that 152 is nearest to.**

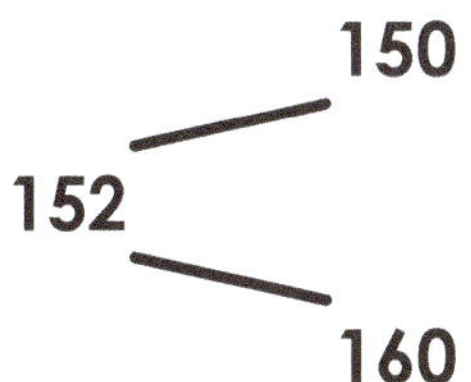

2. Round 137 to the nearest 10.

a. Find 137 on the number line

b. Which two tens does 137 lie between?

c. Circle the ten that 137 is nearest to.

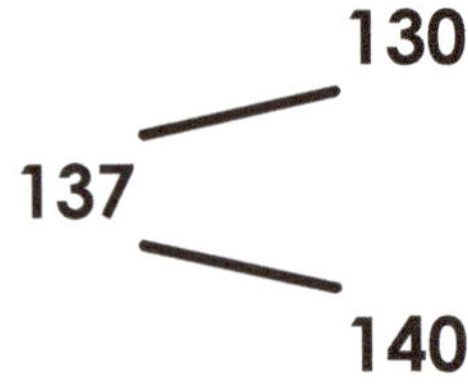

3. Circle the ten that each of the following is nearest to.

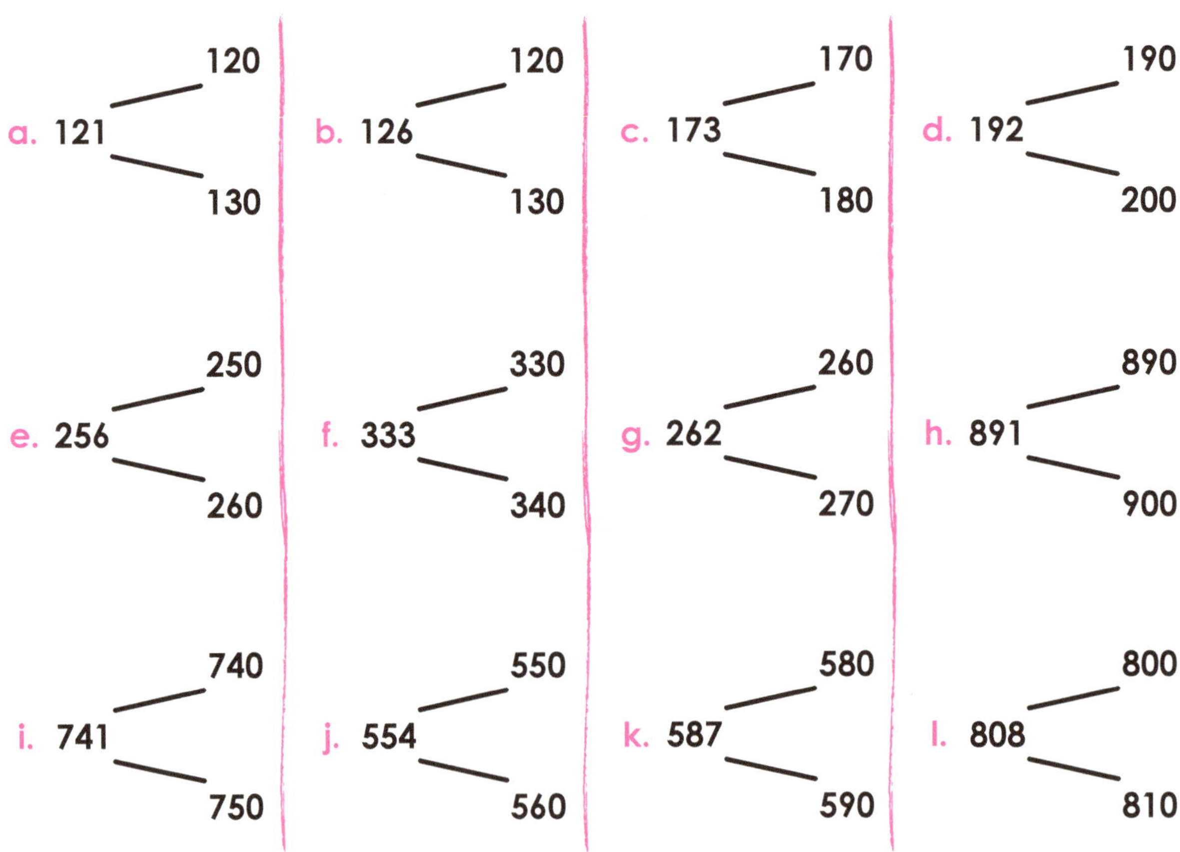

Round the following to the nearest 10

The first one has been done for you.

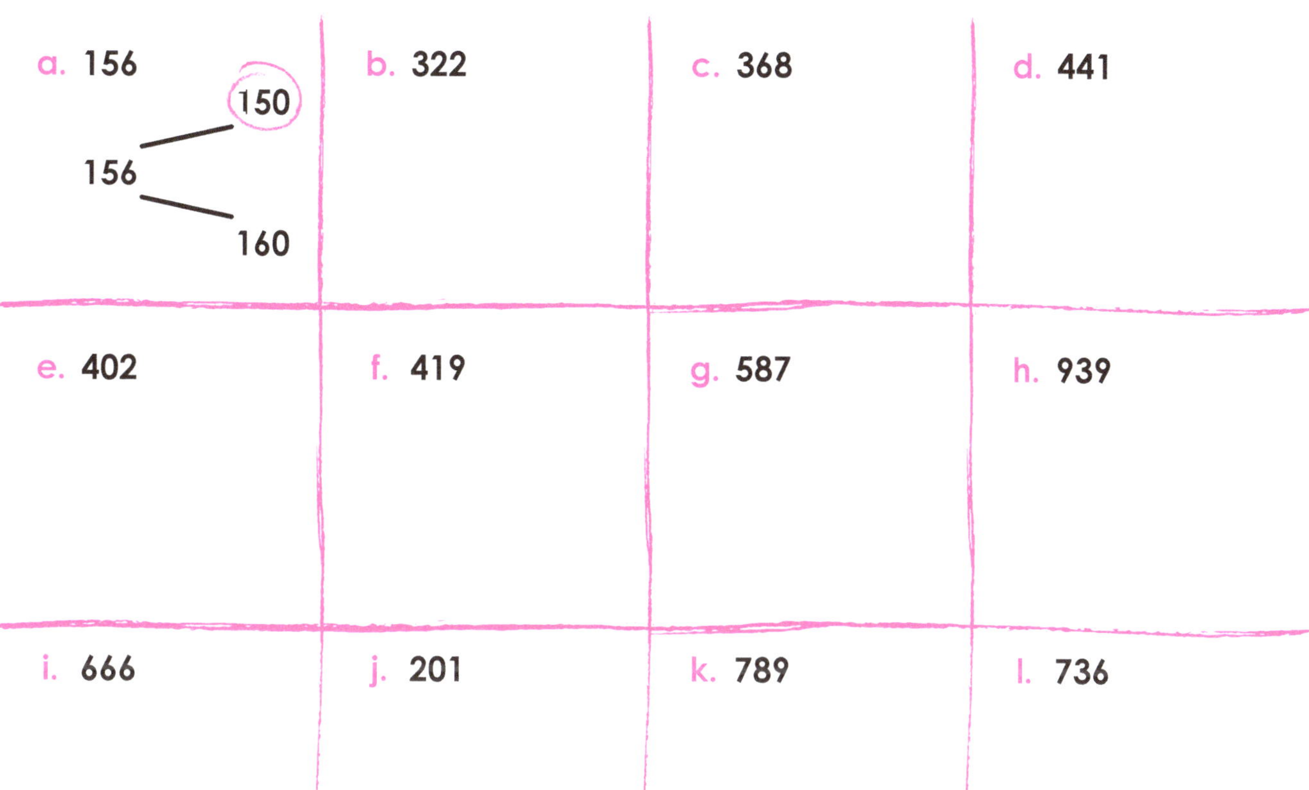

a. 156 → 156 → 150 (circled) / 160	b. 322	c. 368	d. 441
e. 402	f. 419	g. 587	h. 939
i. 666	j. 201	k. 789	l. 736

Round the following amounts to the nearest 10.

a. 561p .... b. £4.32 .... c. £5.54 .... d. £8.91 ....

e. £7.04 .... f. £3.01 .... g. £9.53 .... h. £8.88 ....

Round the following measurements to the nearest 10.

a. 5.64m .... b. 4m32cm .... c. 9.89m .... d. 2.32m ....

e. 5m44cm .... f. 4m56cm .... g. 9m99cm .... h. 6m9cm ....

Problems

1. Jane bought a magazine costing £1.52 and a packet of sweets costing 42p.

a. Round each value to the nearest 10p.

b. Estimate the total cost of the sweets.

c. Find the exact value that Jane had to pay.

2. Mr Jones bought three planks of wood. They measured 4m59cm, 2m,6cm and 3m42cm.

a. Round each length of wood to the nearest 10cm.

b. Estimate the total length of the wood.

c. Find the exact length of the wood.

3. Mary bought a pencil costing 62p and a sharpener costing 53p.

a. Round each value to the nearest 10p.

b. Estimate how much Mary had to pay.

c. Find the exact cost of the pencil and sharpener.

d. Work out how much change Mary gets from £2.00.

4. Mr. Davis is building a wall. His wall is 10 bricks long. Each brick measures 33cm.

a. Round the length of brick to the nearest 10cm.

b. Estimate the length of the wall.

c. Will the total length of the wall be more or less than the estimate?

d. Find the total length of the wall.

5. Mrs Green's lounge measures 5.44m long. She wants to fit a wall unit measuring 2m64cm and a sofa measuring 2m78cm along the wall.

a. Round the length of the wall unit and the sofa to the nearest 10cm.

b. Estimate the total length of the wall unit and the sofa.

c. Using the estimates, can you tell if the wall unit and the sofa will fit along the wall?

d. Find the exact length of the wall unit and the sofa.

e. Will the wall unit and the sofa fit along the wall? Explain your answer.

Notes

Notes

153

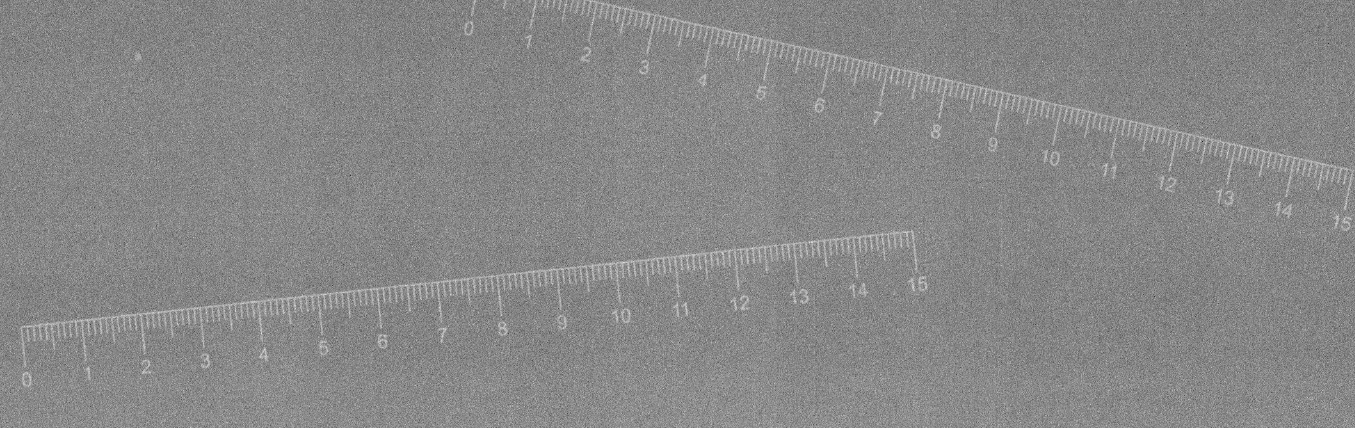

Unit 9

Using a Ruler

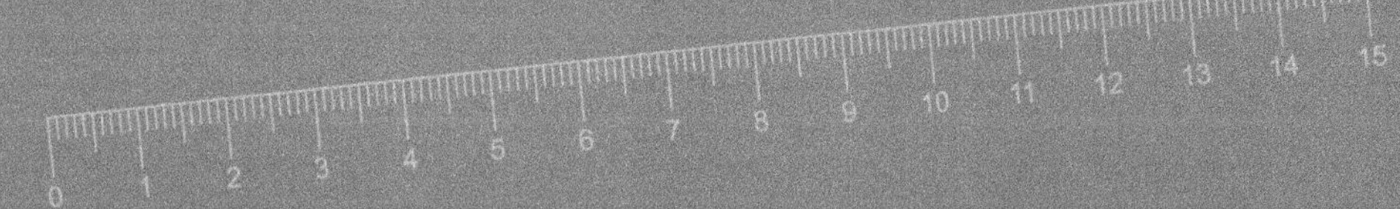

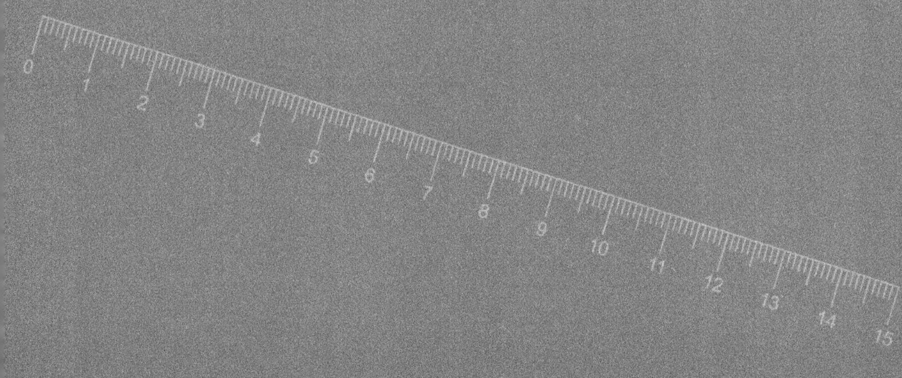

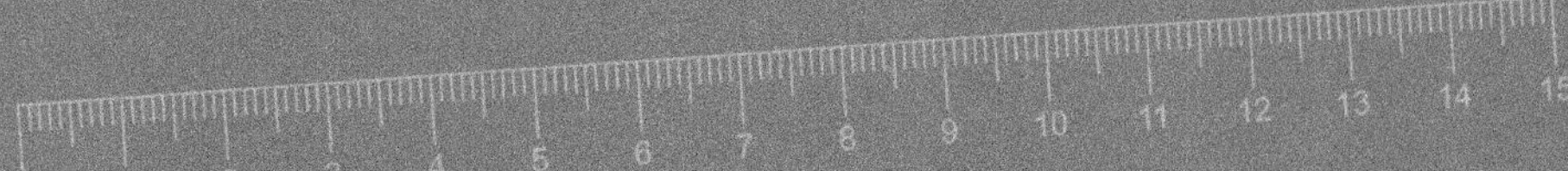

Teacher Note

Use of ruler for measuring and drawing lines

Outcome: Number, money and measurement

Strand: Measure and estimate

Target: Measure in easily handled standard units.

Level: B

Introduction

All pupils should have access to the use of a ruler at least 15cm long. The use of the ruler should be discussed. Pupils should be told where on the ruler to start measuring from, and how to hold the ruler.

Measurement – Using a ruler

Sometimes we measure LENGTHS using a ruler.
Here is a ruler. One side is marked into units called

CENTIMETRES or cm for short.

Start measuring here

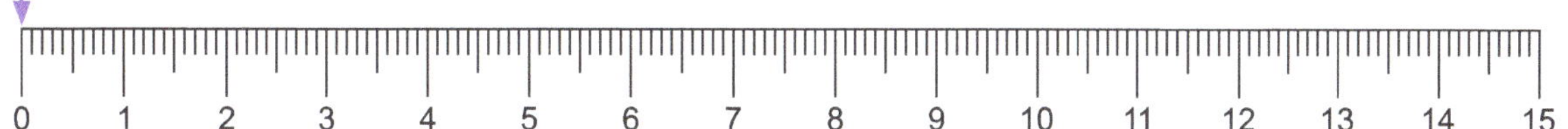

1. Up to how many centimetres can this ruler measure?

2. Measure this line   The line measures 2cm

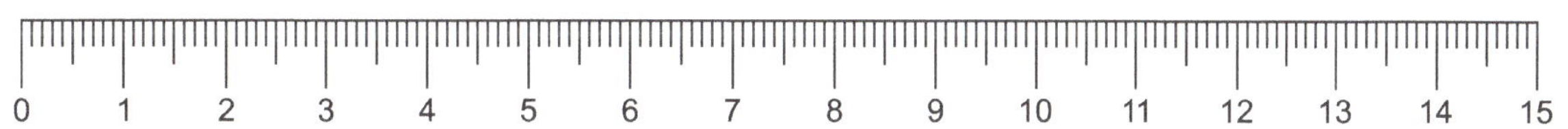

3. Measure this line   The line measures....cm

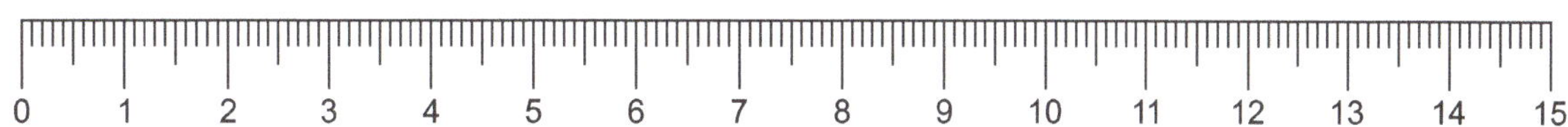

4. Measure this line   The line measures....cm

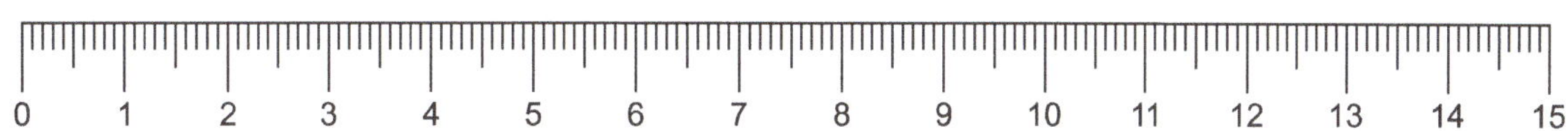

Now...

Use your ruler to measure the lengths of the lines below.
Include the unit cm in your answers.

5. a. ___   The line measures.........................

    b. __________   The line measures........................

    c.   The line measures.........................

                                                        d.
                    The line measures........................

    e. _______________________
       The line measures........................

    f. ____
       The line measures........................

    g. _______________________
       The line measures........................

Here is a shape

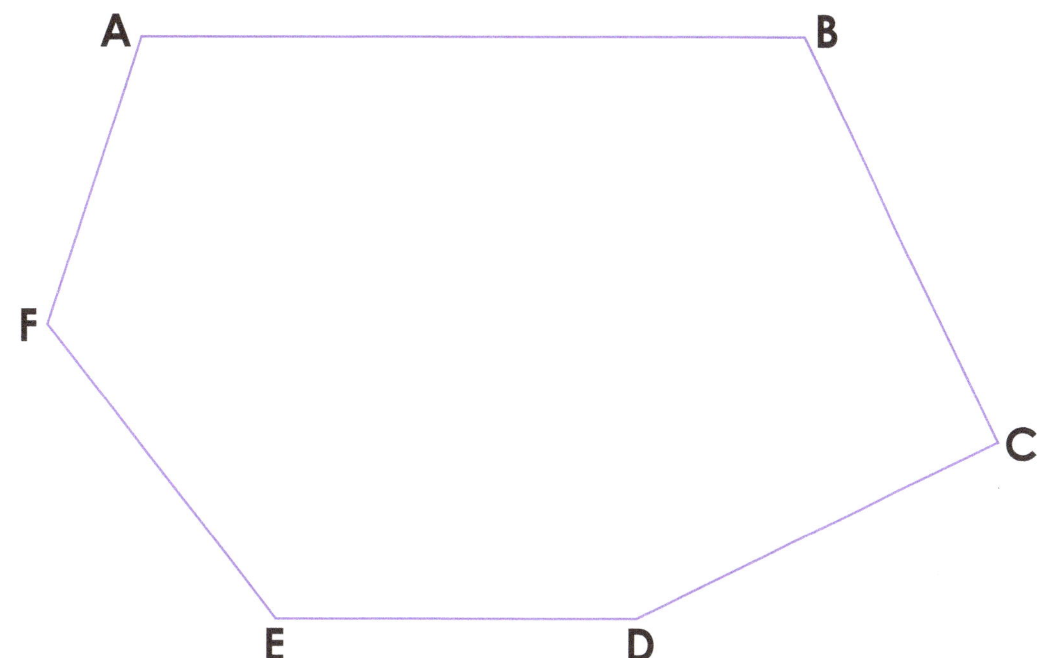

6. Measure the lengths of the sides named below.

   AB = ...   BC = ...   CD = ...

   DE = ...   EF = ...

7. Find the lengths of the sides of the shape below.

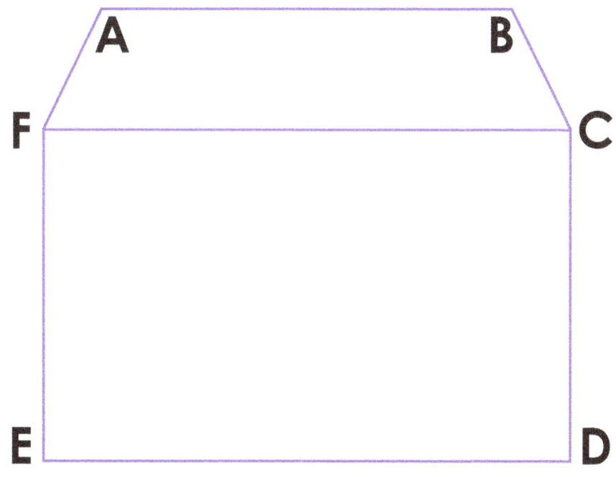

Copy and complete

AB = ...
BC = ...
CD = ...
DE = ...
EF = ...
FC = ...
FA = ...

Problems

8. **Guess** the lengths of the lines below and write your guess down for each.

a. ⎯

The line measures ....

b.

The line measures ....

c.

The line measures ....

d. The line measures ....

e.

The line measures ....

Now...

Check your guess for each line by measuring the length of each line with a ruler. Write the measurement beside your guess for each

9. Draw lines measuring:

   a. 2cm   b. 4cm   c. 10cm   d. 7cm

   a. 1cm   b. 3cm   c. 6cm   d. 12cm

10. Draw a square of side 3cm.

11. Draw a square of side 7cm.

12. Draw a rectangle whose short sides measure 2cm and
    long sides measure 5cm

Finally

Place the following lengths in order starting with the smallest:

1. 5cm   3cm   10cm

2. 1cm   12cm   4cm   6cm

3. 2cm   9cm   8cm   1cm

4. 11cm   5cm   3cm   13cm

Place the following lengths in order starting with the largest:

1. 8cm   11cm   2cm

2. 4cm   14cm   44cm

3. 12cm   1cm   5cm   2cm   9cm

Notes

Notes

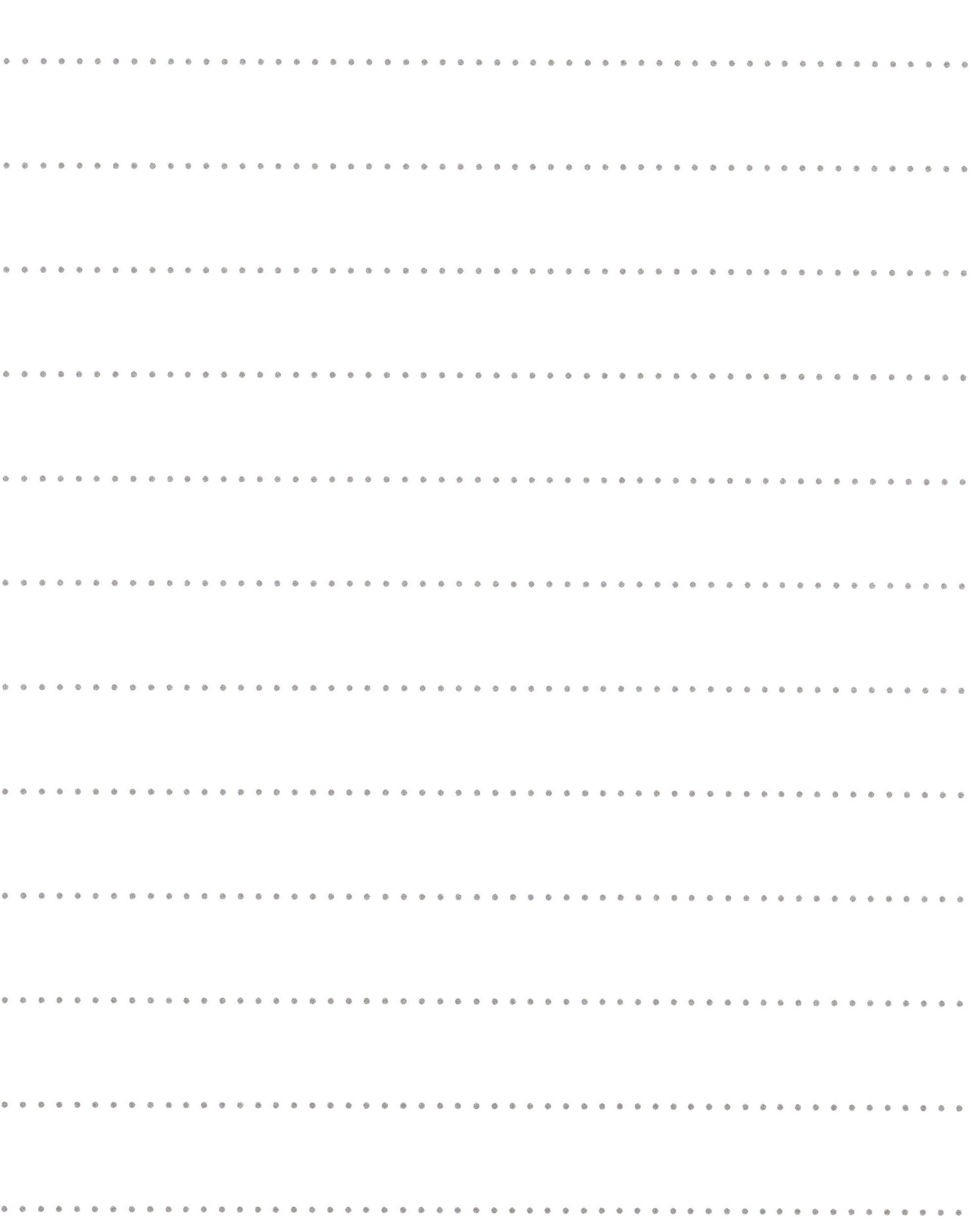

Unit 10

Converting metres to centimetres and Vice versa

Teacher Note

Converting metres to centimetres and vice versa

Outcome: Number, money and measurement

Strand: Measure and estimate

Target: Measure in easily handled standard units –
length: metre, half metre, quarter metre and centimetre.

Level: B

Introduction

The pupils should know that 1m = 100cm, and be able to convert simple measurements from metres to centimetres and vice versa.

Lesson plans for first 3 lessons

Measure the length of a desk using handspans, to the nearest whole handspan.

Pupils should discover that different people get different answers, as handspans vary.

This leads to the introduction of a standard unit of length – the metre

Use the metre stick to measure various things in the classroom/school:

a. to the nearest m

b. to the nearest 1/2 m

c. to the nearest 1/4 m

d. to the nearest cm

Examples

Measure classroom to the nearest m

Measure corridors to the nearest 1/2 m

Measure dimentions of playground to the nearest m

Measure tables to nearest cm

Now...

Make plans of the playground/classroom

Measurements ....

....

....

....

....

Metres and centimetres

Look at a metre stick. (Your teacher will show you one.)

How many centimetres are there in a metre

cm = centimetre

m = metre

1 metre = 100 centimetres

Instead of writing this, a shorter way is:

m = metre

cm = centimetre

1 m = 100 cm

How many centimetres are there in

a. 1/2 m   ....   b. 1/4 metre   ....

a. 1/2 metre   ....   b. 1/4 m   ....

Which answers above are equal?

Metres and centimetres

1. How many centimetres are these in

   a. 1 metre ....   b. 3 metres ....   c. 2 metres ....

   d. 5 metre ....   e. 8 metres ....   f. 6 metres ....

2. A dining room measures 4 metres long.
   How many centimetres is this?

3. The dining room is 3 metres wide.
   How many centimetres is this?

4. Here is a picture of Ken's garden.
   It measures 7m long and 2m wide.

   7m
   2m

   What is a. The length in cm?
              b. The width in cm?

5. Here is the plan of Mr Tod's garden.

   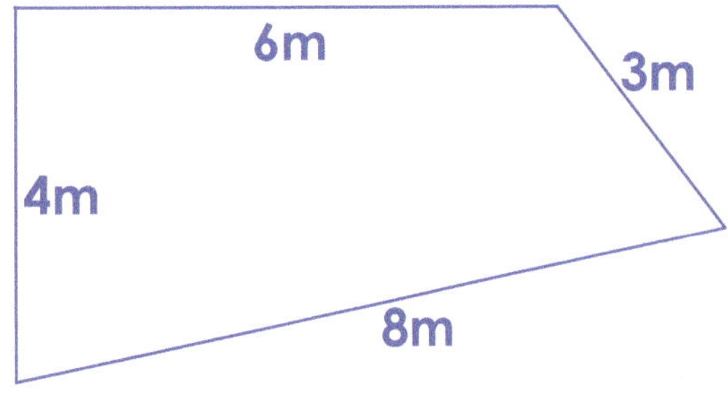

   a. What are the lengths of the sides in cm?

   b. What is the total length of fence that Mr Tod needs to put round his garden?

Finally

6. **How many metres are equal to**

    a. **100 centimetres** ....   b. **200 centimetres** ....

    c. **400 centimetres** ....   d. **600 centimetres** ....

    e. **800 centimetres** ....   f. **900 centimetres** ....

7. **Match the following measurements:**

200 cm   700 cm   500 cm

4 metres   2 m

5 m   400 centimetres   7 metres

Notes

Notes

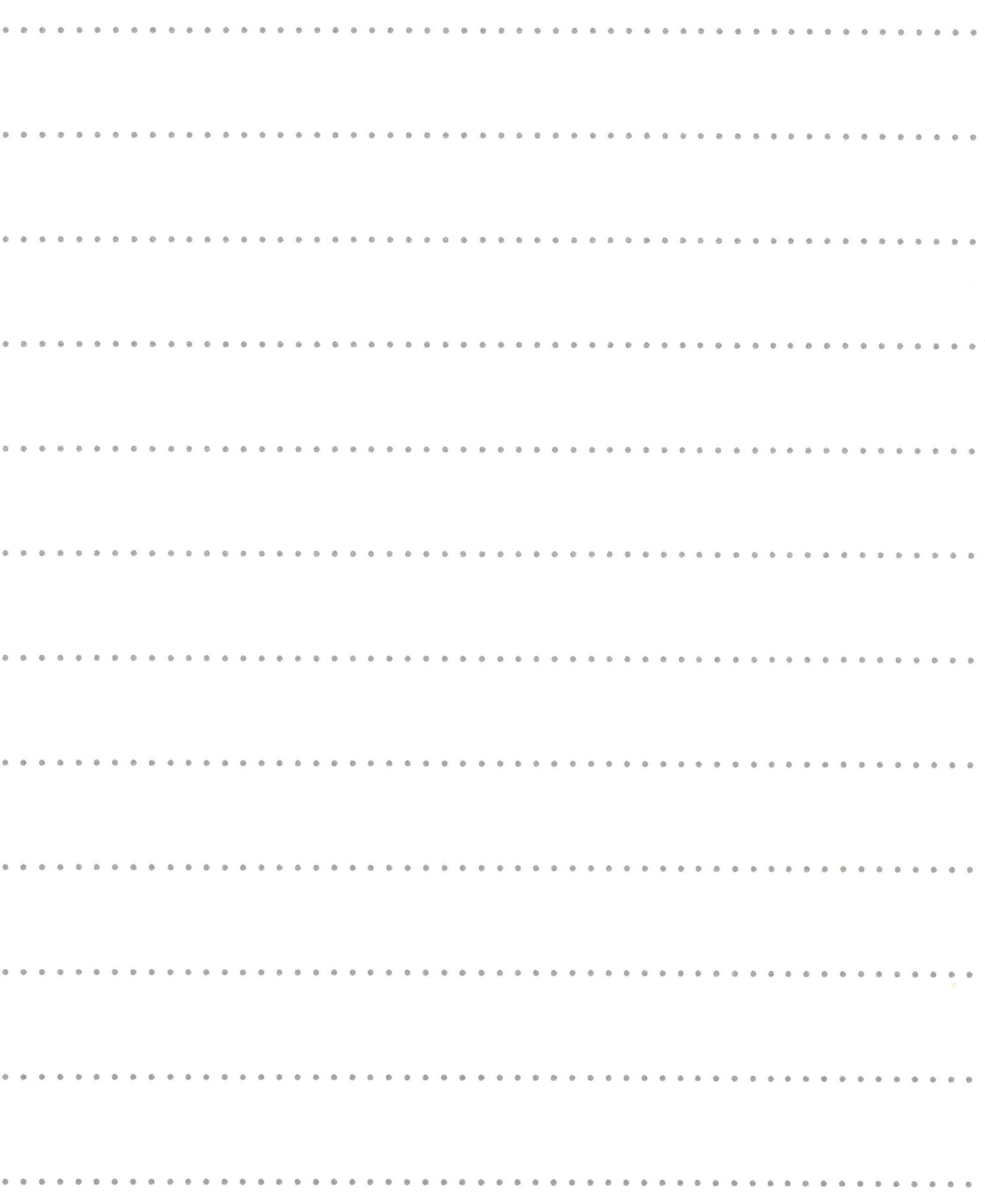

Unit 11

Measurement in standard units

Teacher Note

Measurement in standard units; add and subtract in appropriate measurement. Understand abbreviation; reinforce the relationship between metres and centimetres

Outcome: Number, money and measurement

Strand: Measure and estimate/add and subtract

Target: Measure in standard units.
Add and subtract in applications in measurement.

Level: C

Aims

The pupils should realise the relationship between metres and centimetres, and using the abbreviations m, cm, be able to work with these units.

11

We can write measurements of length in three different ways..

Example 1

Here is a picture of a desk.

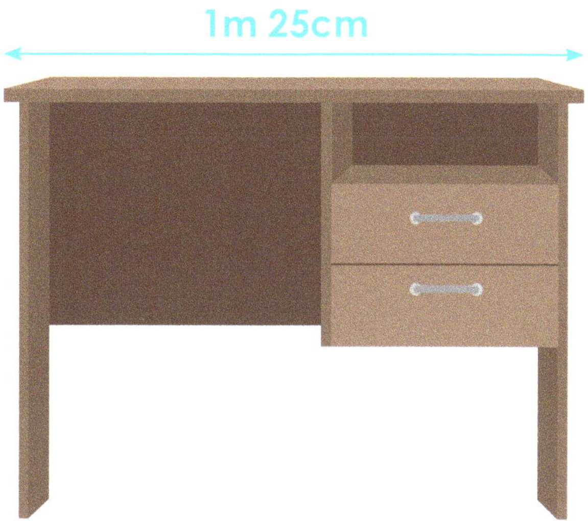

The desk measures	1m 25cm
This is	1 metre 25 centimetres
or	**100cm + 25cm**
	= 125cm

Write the following measurements in 2 more different ways:

1. 1m 50cm

....

....

2. 1m 70cm

....

....

3. 1m 30cm

....

....

4. 1m 40cm

....

....

5. 1m 10cm

....

....

6. 1m 90cm

....

....

Example 2

Here is a picture of a tree

The tree measures   4metres 50 centimetres

This is   4m 50cm

or   400cm + 50cm

= 450cm

11

Write the following measurements in 2 more different ways:

1. 3 metres 20 centimetres
2. 2 metres 20 centimetres
3. 4 metres 25 centimetres
4. 1 metre 50 centimetres
5. 6 metres 30 centimetres
6. 4 metres 90 centimetres
7. 8 metres 80 centimetres
8. 2 metres 25 centimetres
9. 2 metres 50 centimetres
10. 10.5 metres 70 centimetres

Example 3

Here is a picture of a bus

The bus measures 660cm long

This is 600cm + 60cm

= 6m 60cm

or 6metres 60centimetres

Write the following measurements in 2 more different ways:

1. 550cm

2. 320cm

3. 490cm

4. 225cm

5. 440cm

6. 280cm

7. 880cm

8. 940cm

9. 150cm

10. 125cm

11. 160cm

12. 230cm

Adding lengths

Example:　　Add 6m, 3m 20cm

```
      m       cm
      6       00
  +   3       20
      9m      20cm
```

1. Add 1m 20cm and 2 metres 60cm

2. Add 3m 50cm and 1m 60cm

3. Add 1m 255cm and 1m 25cm

4. Add 2m 50cm and 2m 25cm

5. Add 4 metres 60 centimetres and 4 metres 60 centimetres

6. Add 9m and 4m

7. Add 630cm and 210cm

8. Add 740cm and 130cm

9. Add 250cm and 250cm

10. Add 100cm and 400cm

Please show how to work out the problem

1. Here is the plan of a pond in Mr Smith's garden. He is going to put edging round the pond. What length of edging does he need?

2. Mrs Davis is making dresses for her 3 daughters. She is going to put ribbon round the waist of each dress. The dresses measure 60cm, 55cm and 50cm round the waist. How much ribbon should Mrs Davis buy?

3. Mrs Jones has a space along her kitchen wall measuring 3 metres 40 centimetres. She wants to fit 4 kitchen units each measuring 90 centimetres wide.

   a. How wide are the kitchen units altogether?

   b. Can Mrs Jones fit 4 units into the space?

Problems

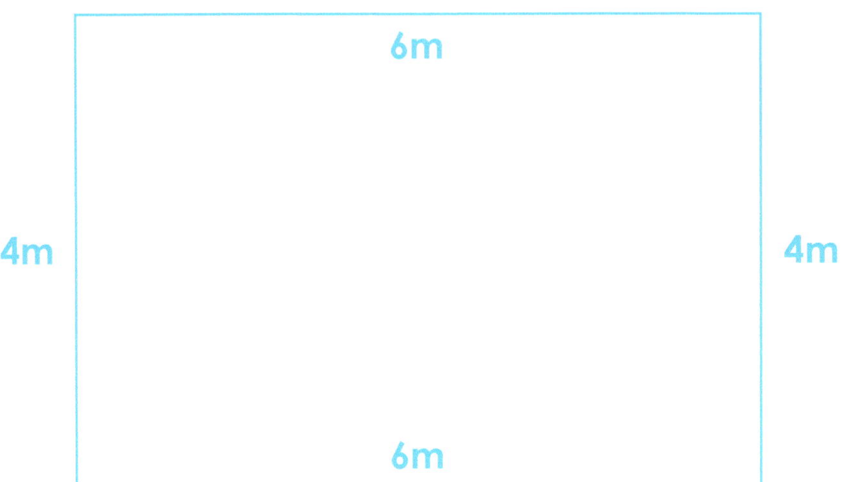

4. Here is the plan of Mr Jones lawn. He is going to buy small trees to plant round the border 1m apart. How many trees should Mr Jones buy?

5. Here is a picture of Mary's bedroom. The door is 1m wide. Her Dad has decided to put a border around the room. The borders are sold in 10m rolls and cost £3.25

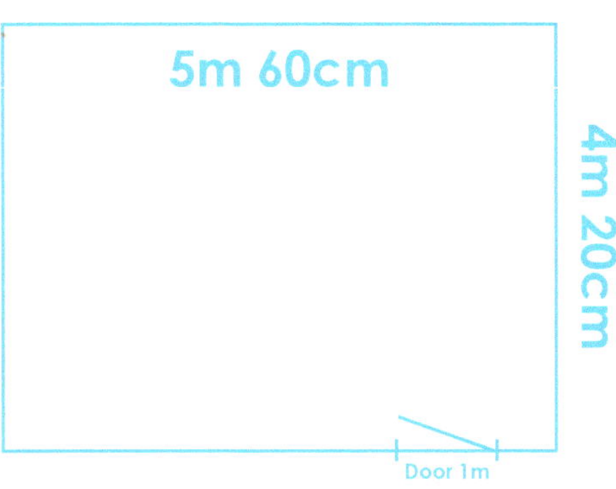

a. Work out exactly the length of border needed to go round the room.

b. How many rolls of border does Dad need to buy?

c. How much do the rolls cost altogether?

Notes

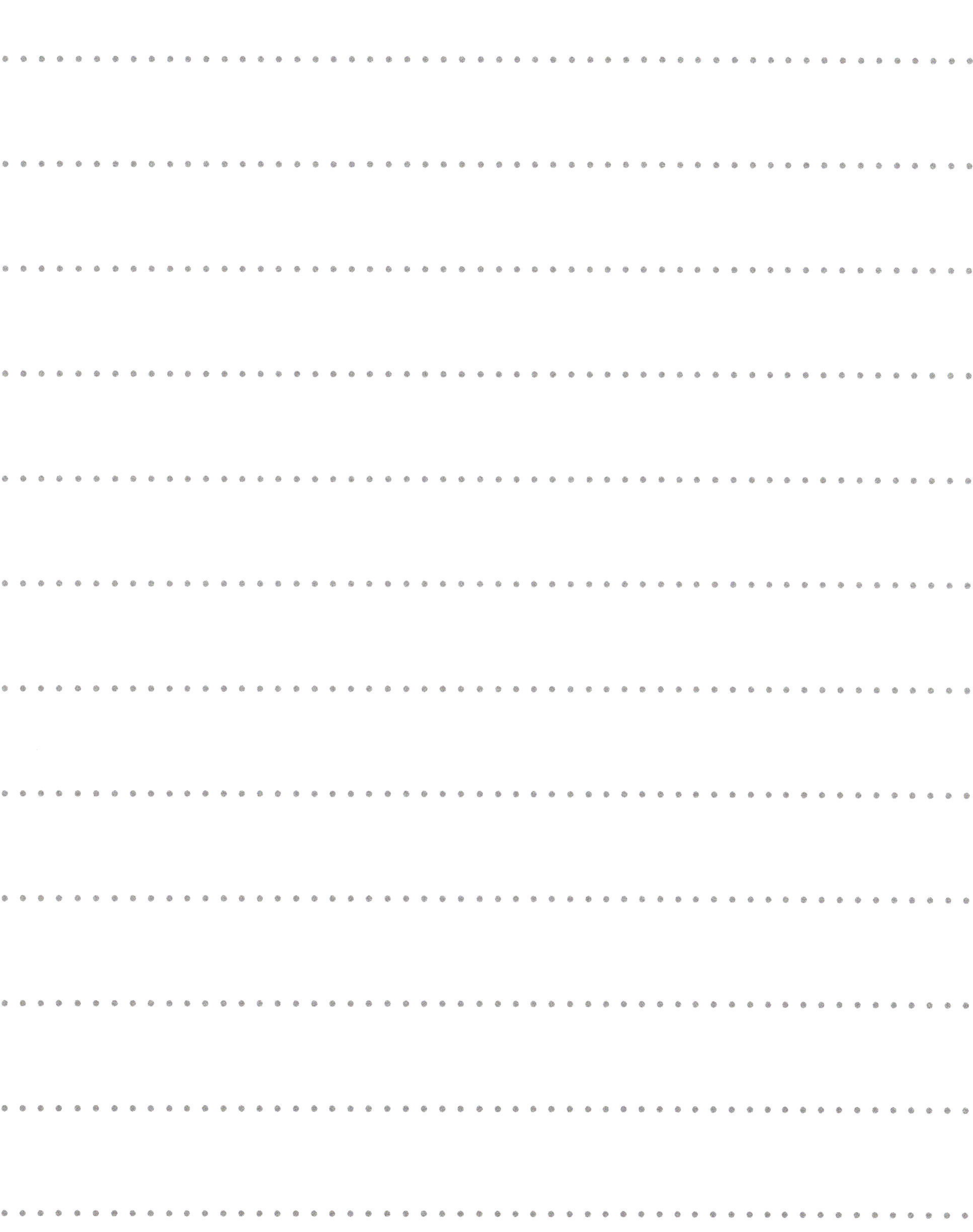

Unit 12

Areas:
square centimetres;
areas and shape

Teacher Note

The pupils should know that areas can be measures in cm^2

Outcome: Number, money and measurement

Strand: Measure and estimate

Target: Measure area in standard units.
Realise that area can be conserved when shape changes.

Level: C

Aims

The pupil should be able to measure areas of shapes in square centimetres.

Areas

Here is a centimetre square.

Its sides measure 1 centimetre

It has an area of 1 square centimetre.

The can be written like 1cm²

Here is another square

How many centimetre squares is it made from?

It has a area of 4 square centimetres

or 4cm²

How many centimetre squares
do each of the shapes below cover?

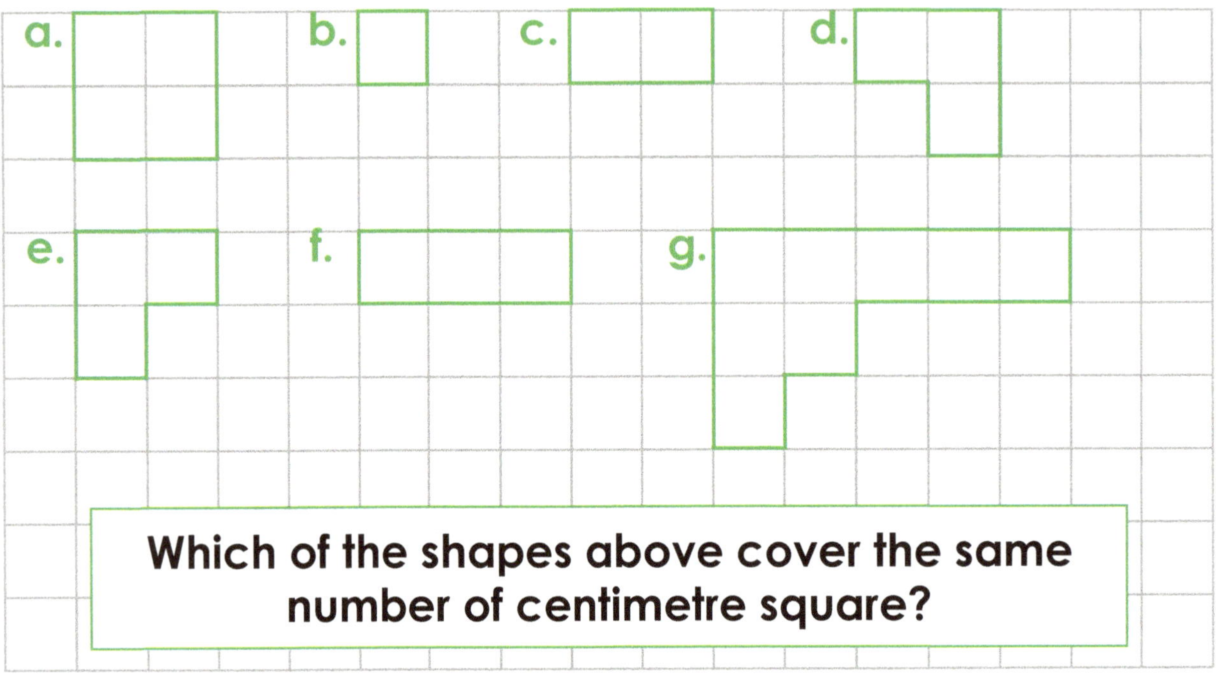

Which of the shapes above cover the same
number of centimetre square?

How many centimetre squares
do each of the shapes below cover?

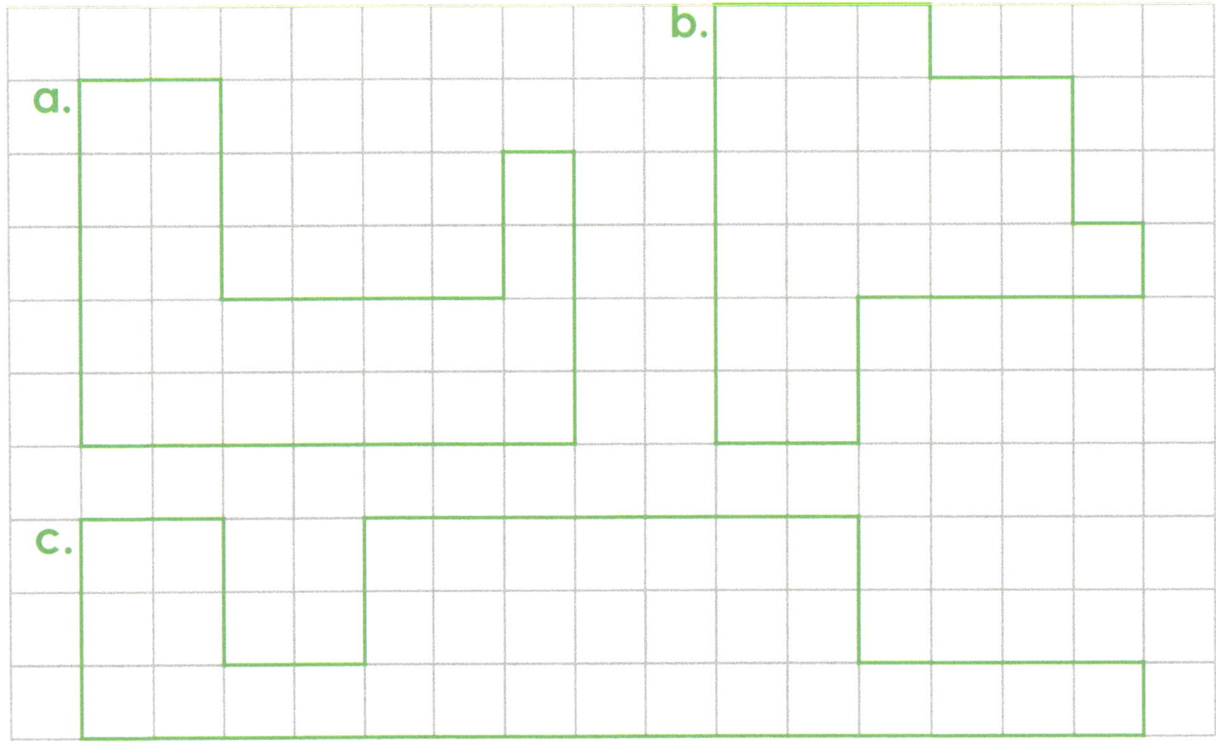

Draw

Draw two shapes covering **4** centimetre squares.

Draw four shapes covering **6** centimetre squares.

Draw eight different shapes each covering **8** centimetre squares. Colour them if you like.

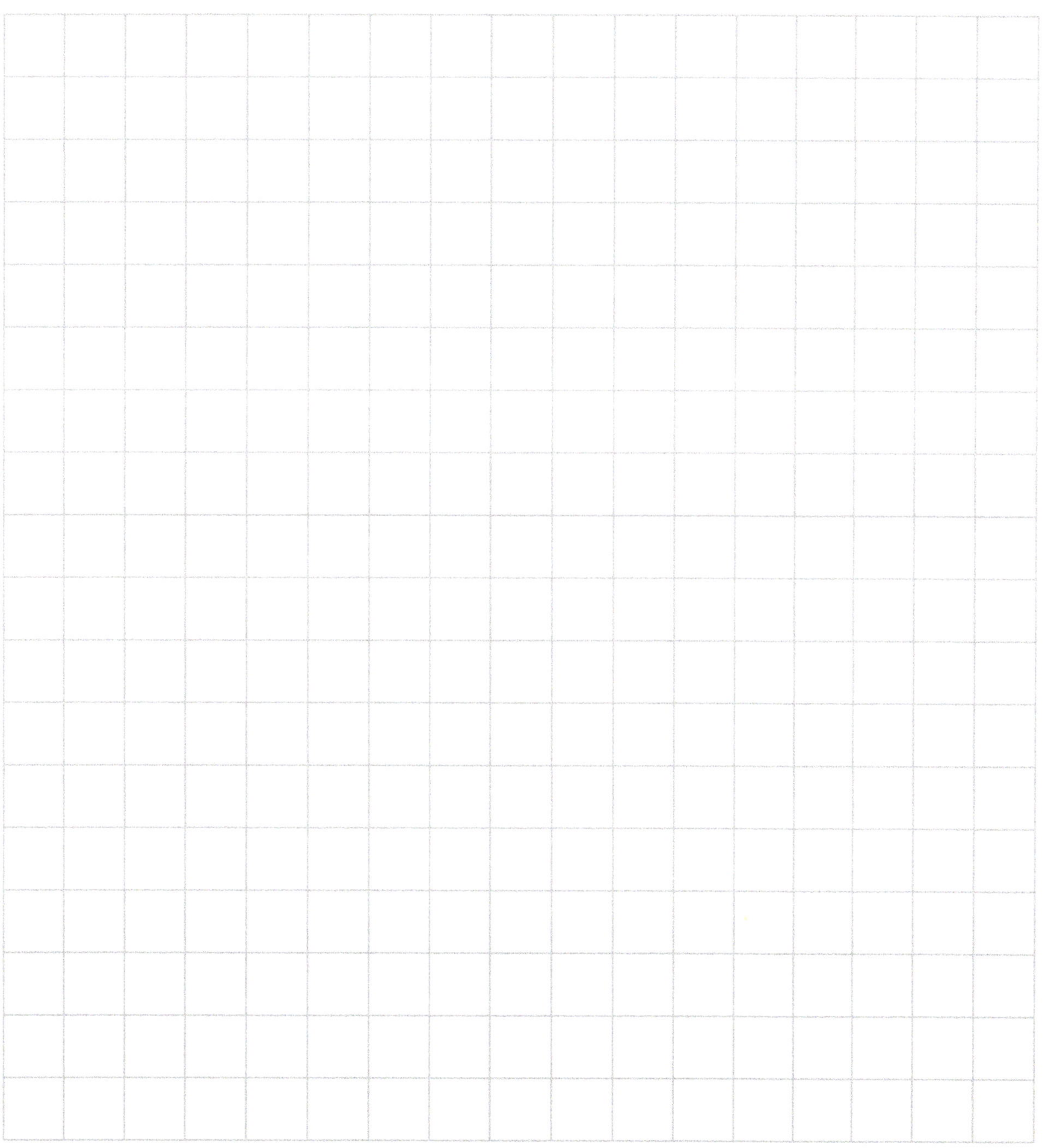

How many centimetre squares do each of the shapes below cover?

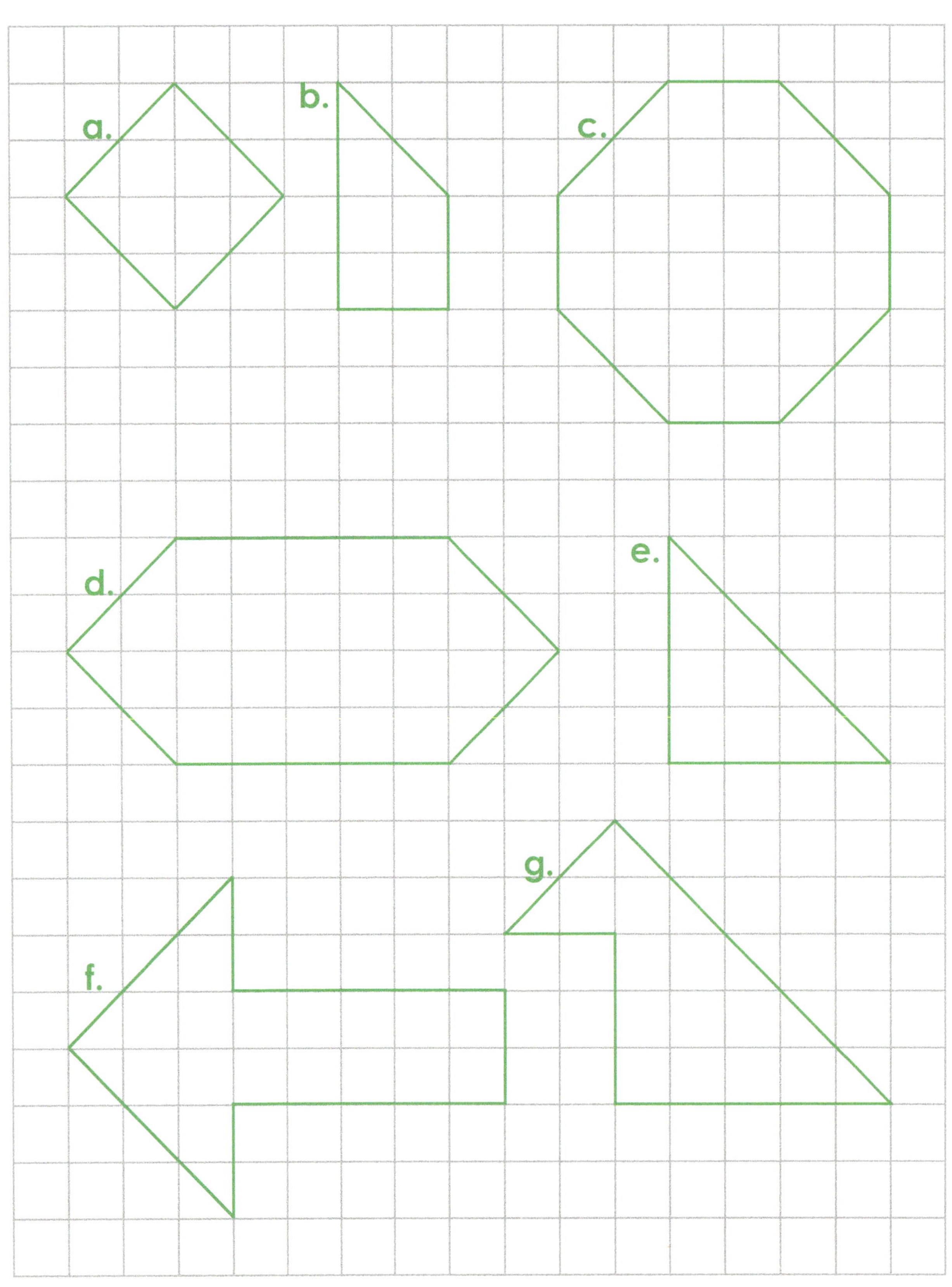

What are the areas of the shapes below in square centimetres? (Write cm² for short)

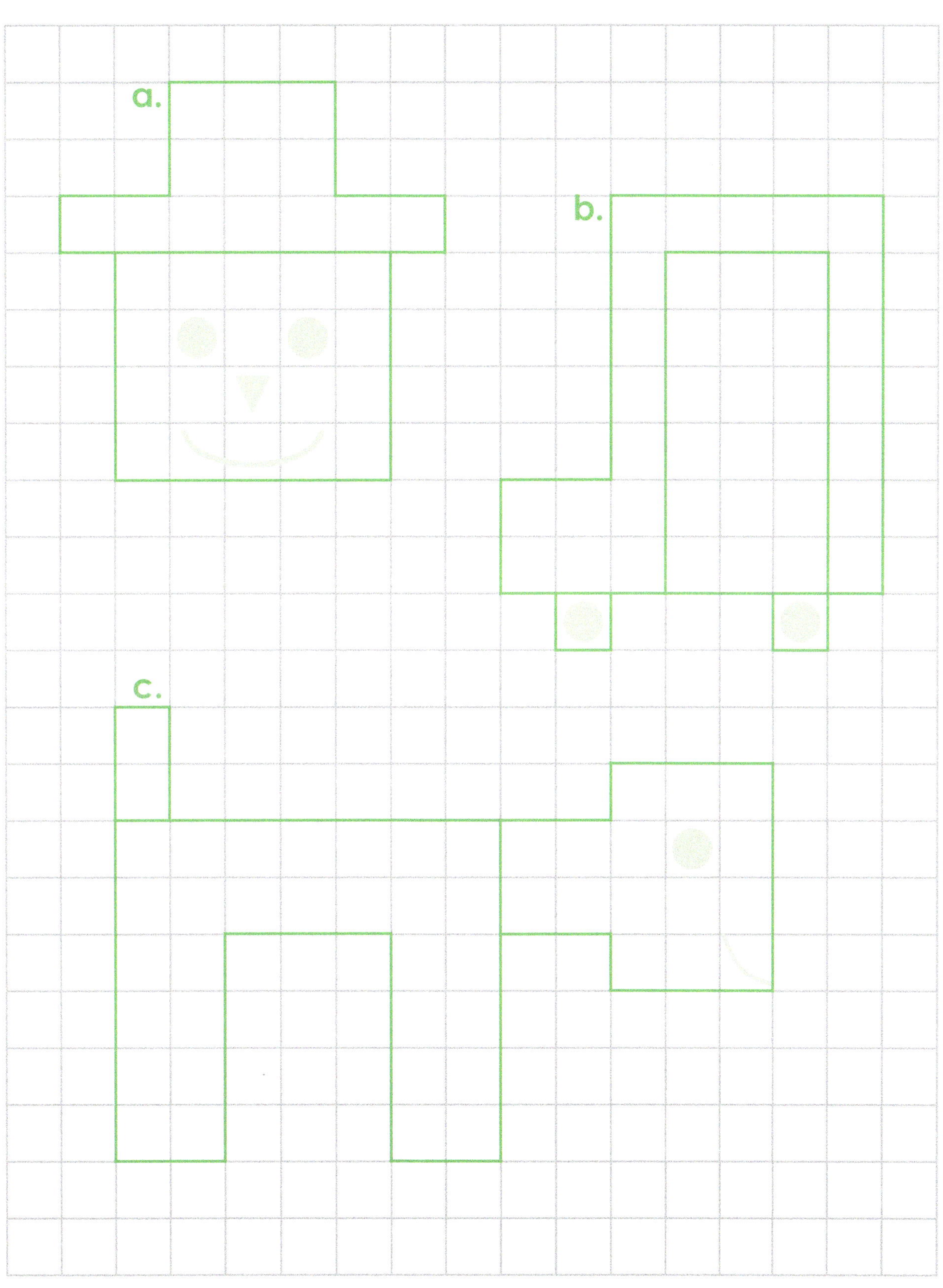

What are the areas of the shapes below in square centimetres? (Write cm² for short)

Draw shapes covering the following areas:
1. 2 cm²   2. 4 cm²   3. 10 cm²   4. 15 cm²

Draw four different shapes covering 6cm²

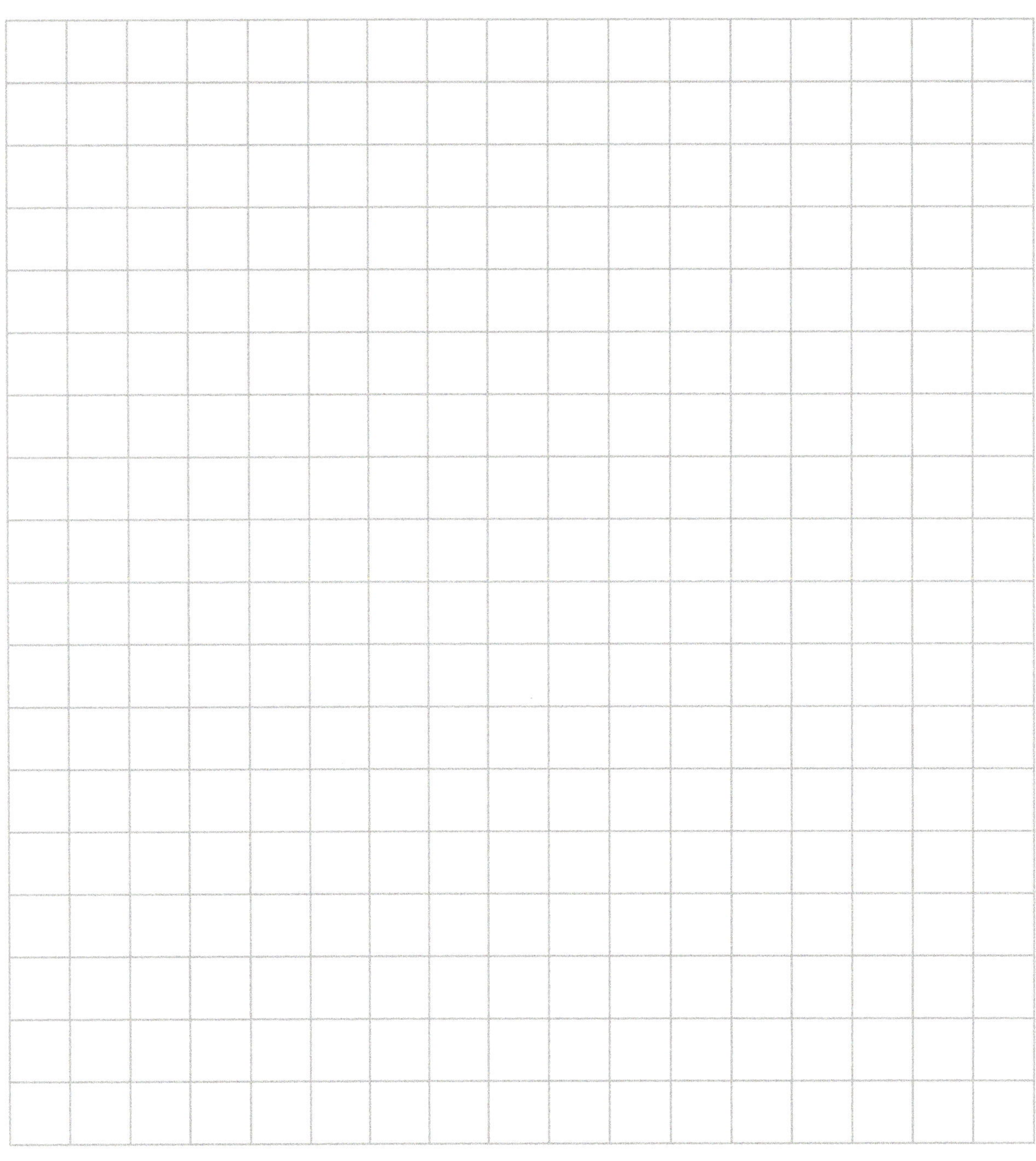

Here is a tile.
How many centimetre square does it cover?
What is its area in cm²?

Make a tiling using this shape on the grid below.

Colour your tiling . What is the area of your tiling?

Here is a tile. What is its area?
Make a tiling using this tile on the grid below.
What is the area of your tiling?

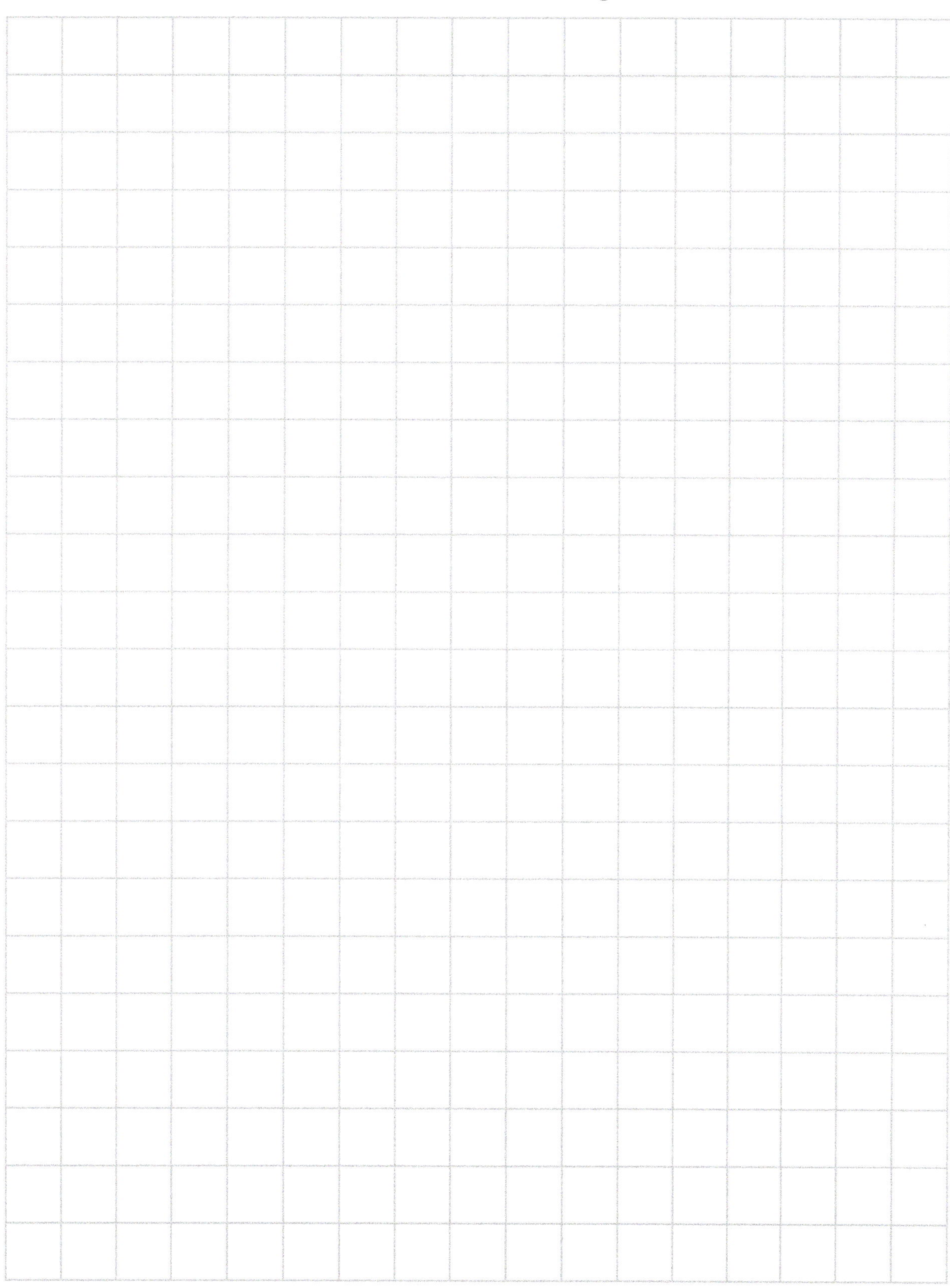

Notes

196

Notes

Unit 13

Functions and Equations

Teacher Note

Functions and equations

Outcome: Number, money and measurement

Strand: Functions and equations

Level: Find the missing numbers in statements where symbols are used for unknown numbers or operators.

Aims

The pupils should understand how to replace a symbol i.e. □, Δ, * by a value or a process when adding, subtracting for values up to 10, multiplying and dividing.

Introduction

Simple examples could be shown and discussed orally and sheet 1 should be read out to pupils. Pupils should be encouraged to set out the questions like this:

$$\square + 3 = 6$$

$$\square = 6$$

Missing values

1. Find the missing value.

$$1 + ? = 3$$

$$? = \,\dots\dots\dots$$

---

2. Find the missing value.

$$2 + ? = 4$$

$$? = \,\dots\dots\dots$$

---

3. Find the missing value.

$$3 + ? = 5$$

$$? = \,\dots\dots\dots$$

13

Problems

1. $1 + \square = 2$

2. $4 + \square = 6$

3. $3 + \blacktriangle = 4$

4. $5 + \square = 9$

5. $4 + \square = 8$

6. $9 + \square = 10$

7. $8 + \blacktriangle = 9$

8. $7 + \square = 10$

9. $4 + * = 8$

10. $2 + * = 4$

11. $\bullet + 1 = 3$

12. $* + 2 = 5$

13. $\blacktriangle + 4 = 9$

14. $\square + 3 = 7$

15. $\square + 2 = 8$

16. $* + 5 = 9$

17. $* + 2 = 6$

18. $\square + 9 = 10$

19. $\blacktriangle + 3 = 6$

20. $\blacktriangle + 4 = 7$

Problems continued

21. □ - 1 = 1

22. □ - 3 = 4

23. ▲ - 1 = 4

24. □ - 2 = 3

25. ▲ - 3 = 6

26. ✳ - 1 = 9

27. ✳ - 2 = 8

28. ▲ - 2 = 5

29. □ - 3 = 7

30. ▲ - 4 = 5

31. ▲ - 2 = 5

32. □ - 2 = 7

33. ✳ - 4 = 2

34. ▲ - 4 = 1

35. □ - 5 = 3

36. ▲ - 3 = 5

37. 9 - □ = 8

38. 8 - ▲ = 3

39. 5 - □ = 1

40. 4 - ▲ = 3

Find the missing values

1. $2 \times \square = 2$

2. $3 \times \square = 3$

3. $3 \times \square = 6$

4. $2 \times \square = 10$

5. $2 \times \blacktriangle = 8$

6. $5 \times \blacktriangle = 25$

7. $10 \times \ast = 40$

8. $2 \times \square = 6$

9. $5 \times \square = 45$

10. $10 \times \square = 20$

11. $\square \times 2 = 4$

12. $\blacktriangle \times 2 = 10$

13. $\square \times 5 = 30$

14. $\blacktriangle \times 10 = 30$

15. $\blacktriangle \times 1 = 4$

16. $\ast \times 3 = 9$

17. $\blacktriangle \times 3 = 15$

18. $\square \times 9 = 18$

19. $\blacktriangle \times 2 = 10$

20. $6 \times \square = 60$

13

Find the missing values

1. □ ÷ 2 = 2

2. ▲ ÷ 2 = 5

3. □ ÷ 2 = 6

4. ✳ ÷ 2 = 8

5. ▲ ÷ 2 = 10

6. ✳ ÷ 2 = 4

7. □ ÷ 2 = 1

8. ✳ ÷ 2 = 3

9. □ ÷ 2 = 7

10. □ ÷ 2 = 9

Problems

Find **all** the prossible pairs of missing values

---

1. □ + ▲ = 3

---

2. □ + ▲ = 9

---

3. ✳ + □ = 6

---

4. □ + ▲ = 5

---

5. ▲ + □ = 8

Find the processes

1. $1 \square 1 = 2$

2. $2 \square 3 = 5$

3. $2 \square 2 = 4$

4. $5 \square 2 = 10$

5. $1 \square 3 = 3$

6. $4 \square 1 = 5$

7. $9 \square 1 = 10$

8. $8 \square 2 = 10$

9. $5 \square 3 = 15$

10. $6 \square 2 = 8$

11. $3 \square 3 = 9$

12. $3 \square 2 = 1$

13. $5 \square 2 = 3$

14. $9 \square 7 = 2$

15. $4 \square 2 = 2$

16. $8 \square 2 = 4$

17. $4 \square 4 = 8$

18. $3 \square 1 = 2$

19. $10 \square 2 = 5$

20. $5 \square 6 = 30$

Make up your own problems

208

Make up your own problems

209

Notes

210

Notes

211

Unit 14

Functions and Equations

Teacher Note

Functions and equations

Outcome: Number, money and measurement

Strand: Functions and equations

Level: Use a simple function machine for operations involving dividing, halving, adding and subtracting.

Aims

So that the pupil can use a simple fraction machine for operations involving adding, subtracting, doubling and halving.

Prior Knowledge

The pupil should know how to add, subtract, multiply and divide for small numbers.

Introduction

An explanation of number machine should be given, with reference to other machines and what can be made using machines. For example, a sweet making machine could be explored. Discussion of what is INPUT, what happens inside the machine and what is OUTPUT is important. Oral examples of the adding and subtracting processes should be discussed with the pupils.

An explanation of what is meant by 'doubling' and 'halving' should be given at the appropriate point in the lesson. Oral examples and written examples should be given and discussed.

Flags
Level C

Here is a number machine:

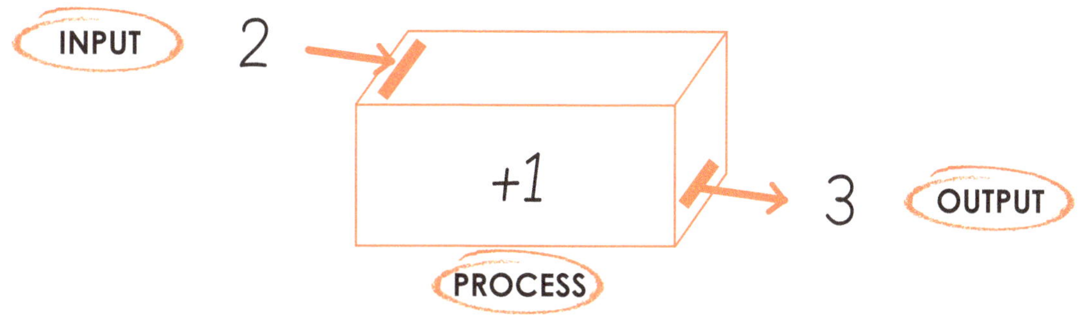

A flag diagram to show this is:

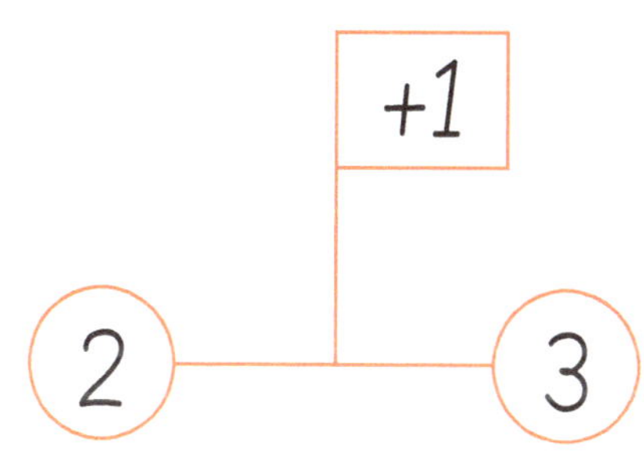

Complete the flag diagrams below to find the output:

1.

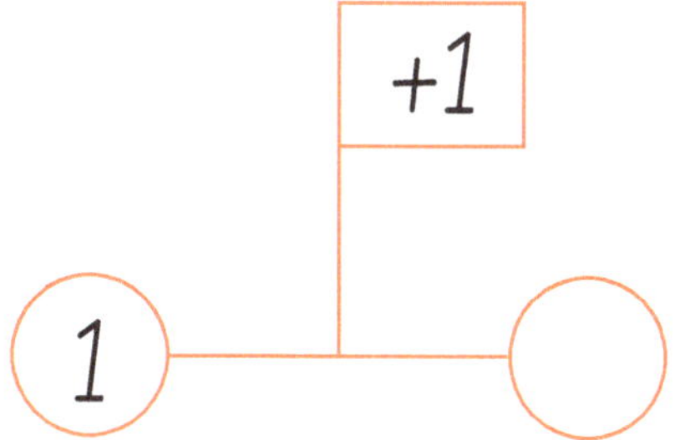

2.

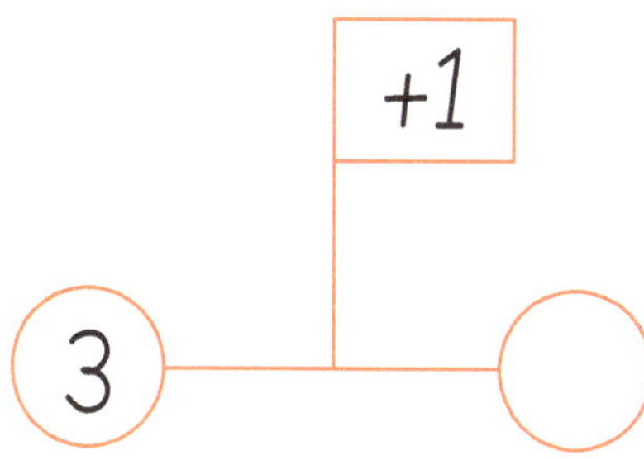

3.

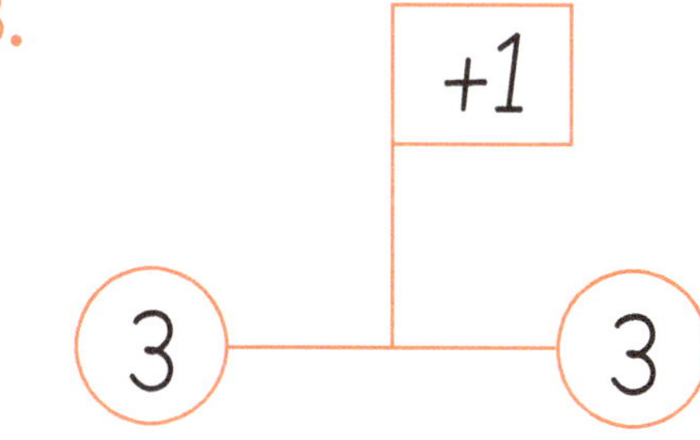

4.

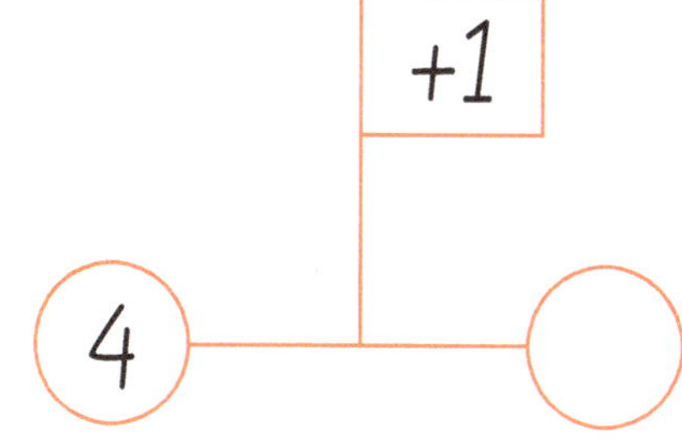

14

Flags
Level C

Find the output for each of the flag diagrams below:

1.
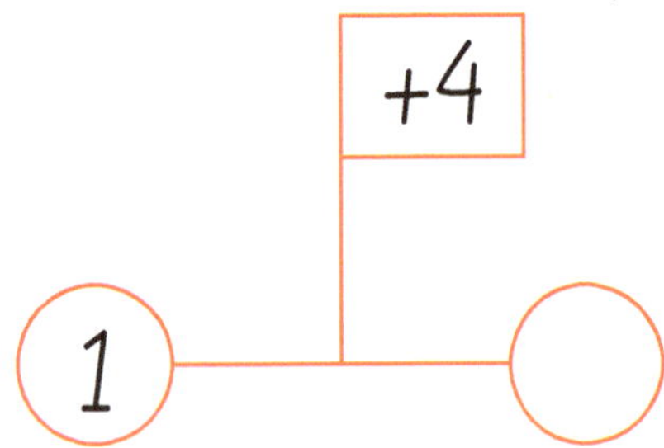

2.
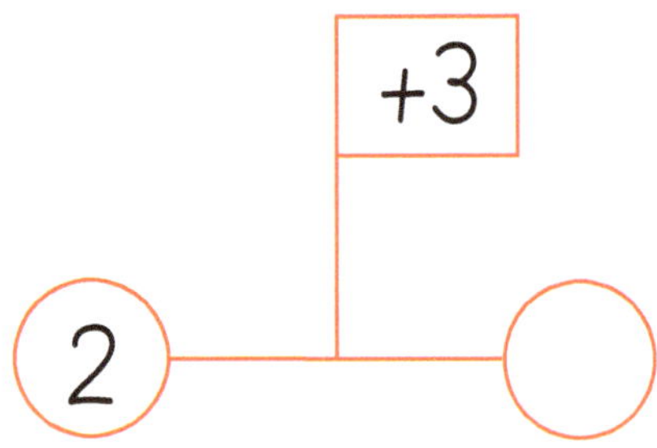

3.
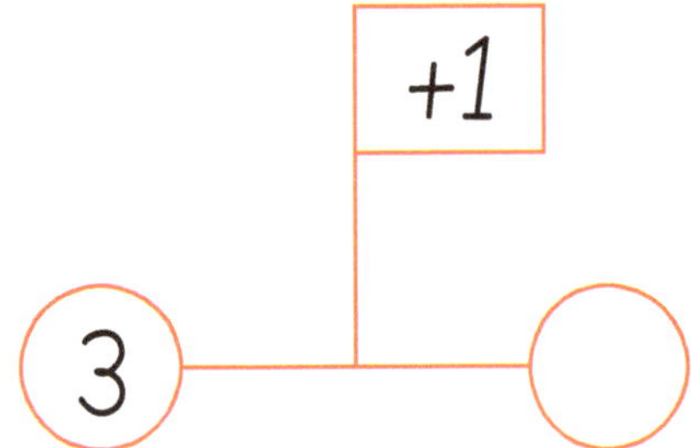

4.
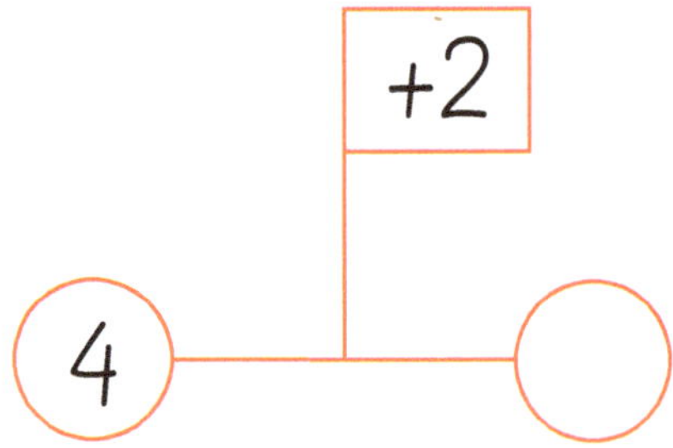

5.
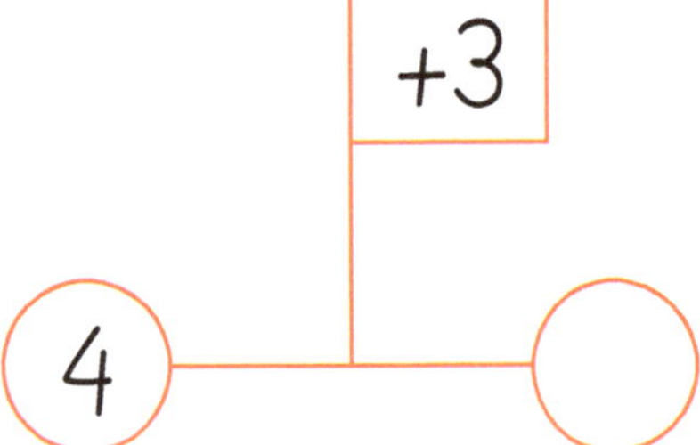

6.
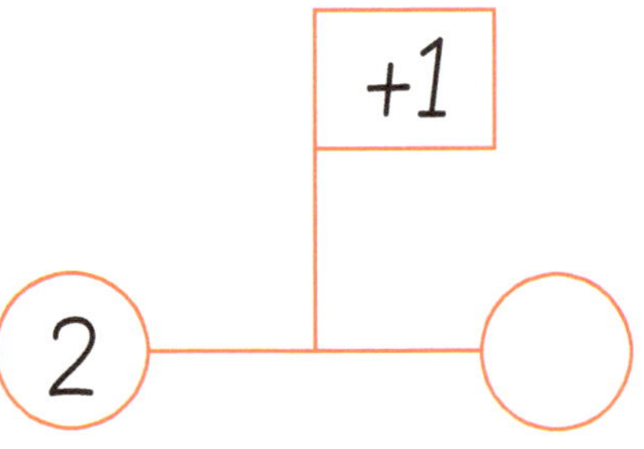

7.
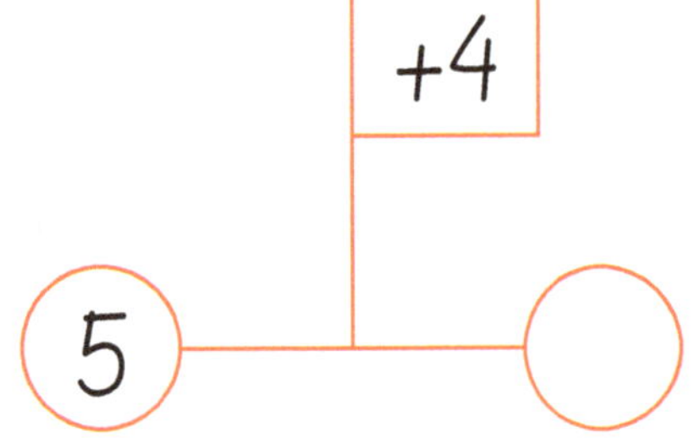

8.
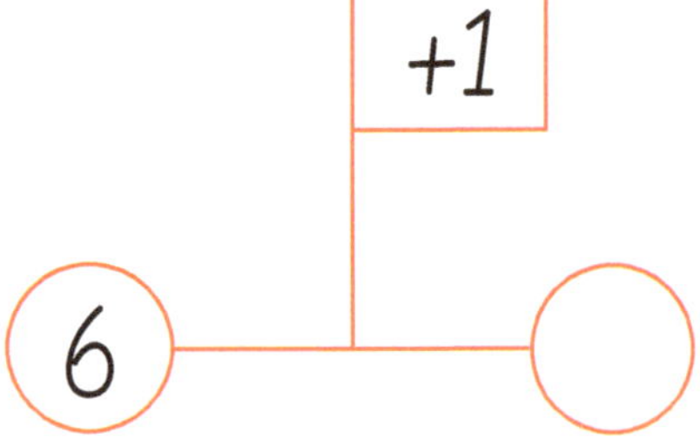

14

Flags
Level C

Find the output for each of the flag diagrams below:

1.
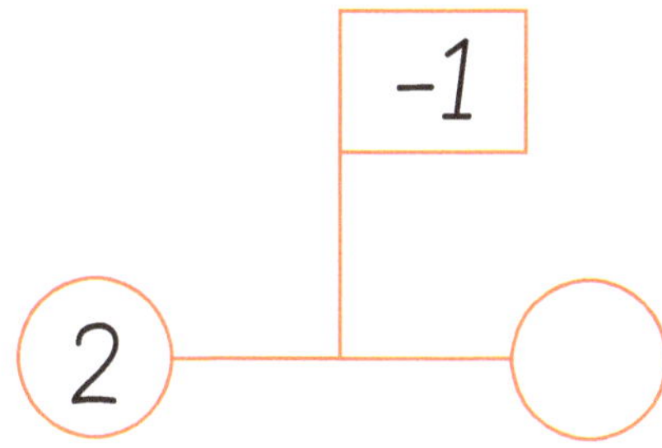

2.
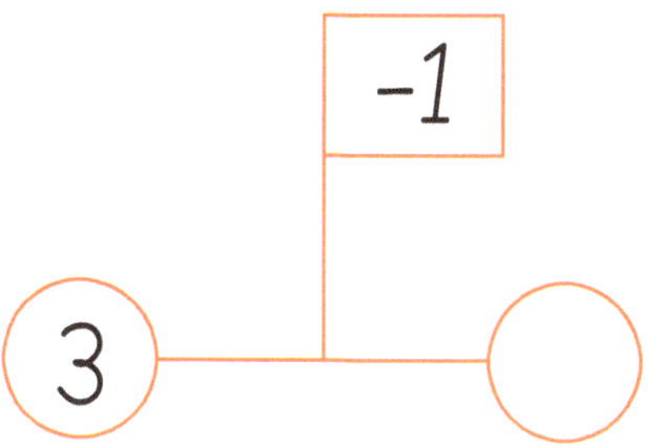

3.
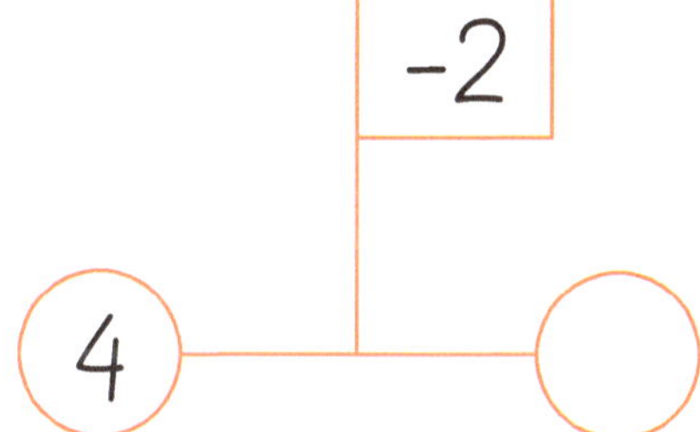

4.
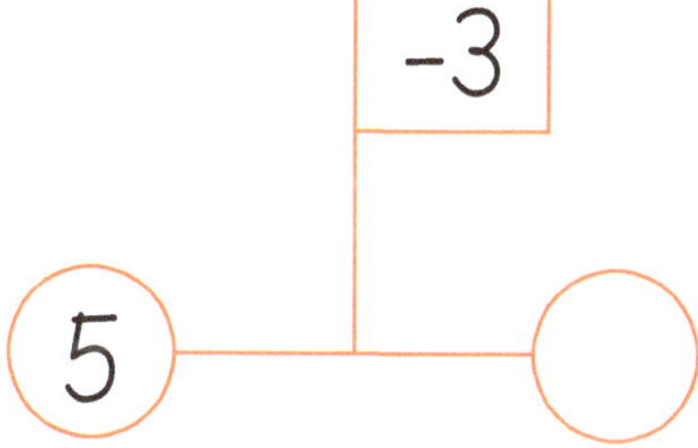

5.
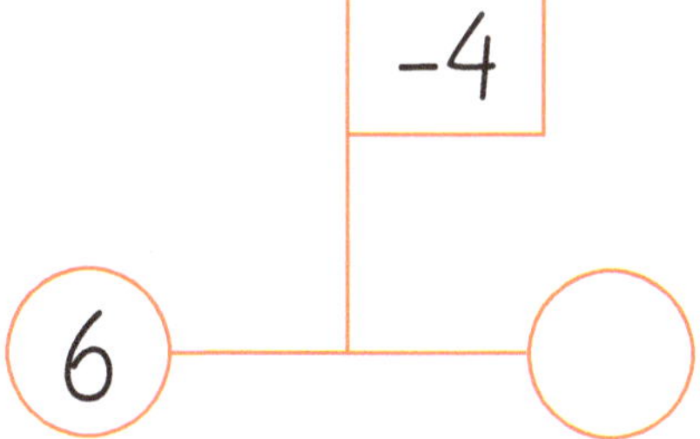

6.

7.

8.
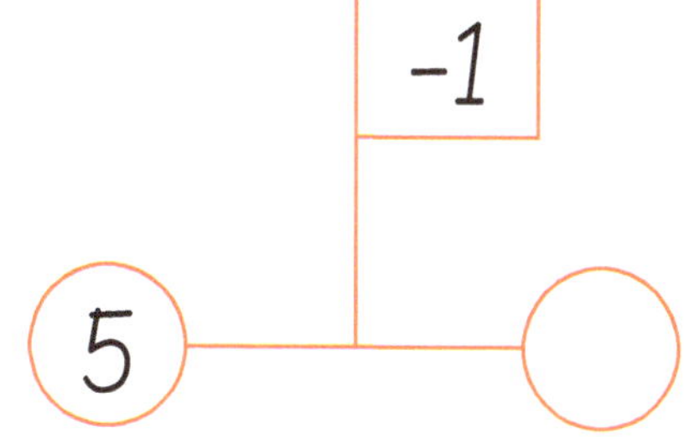

Flags
Level C

Find the output for each of the flag diagrams below:

1.
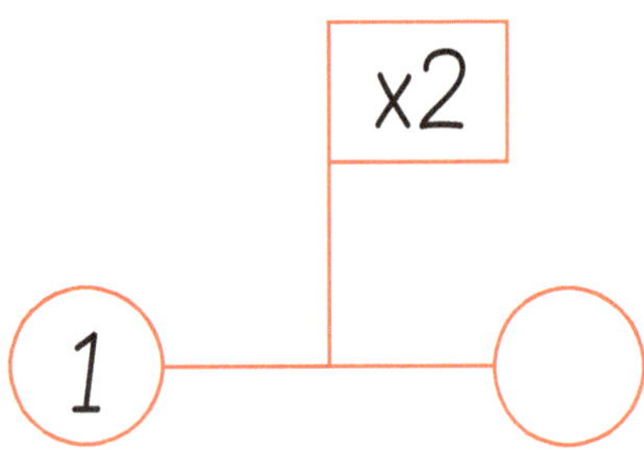

2.
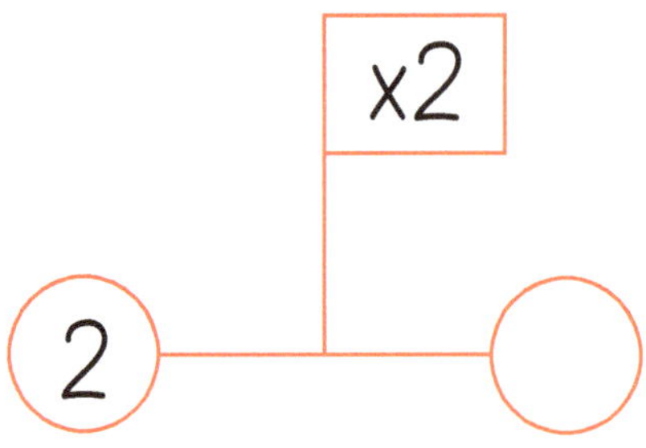

3.
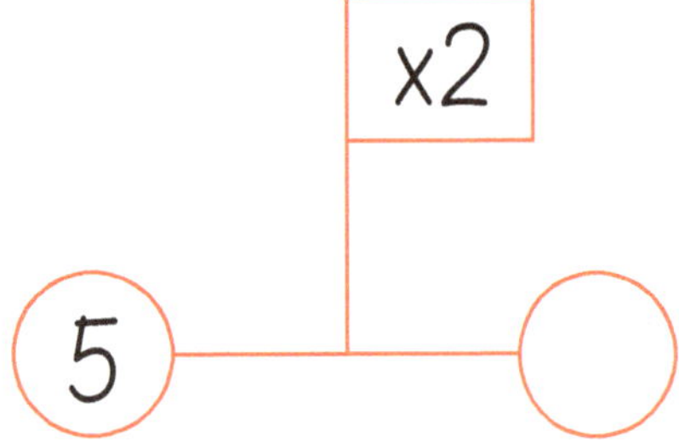

4.
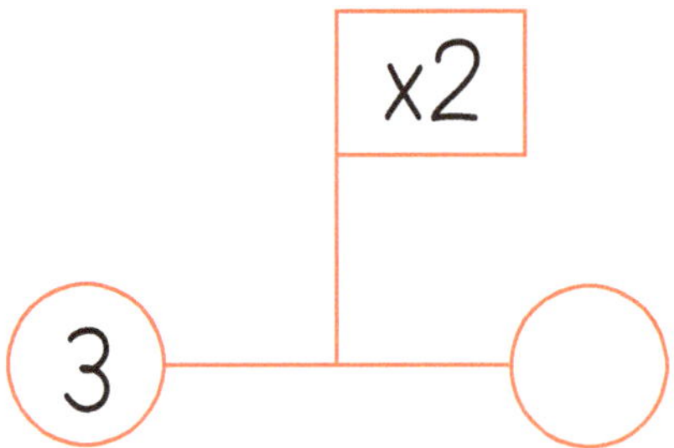

5.
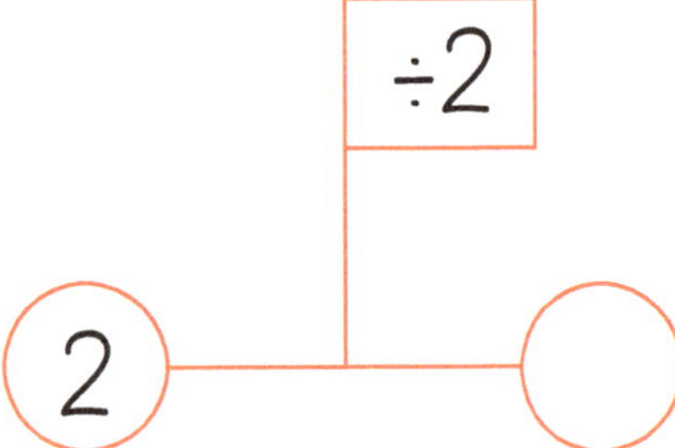

6.
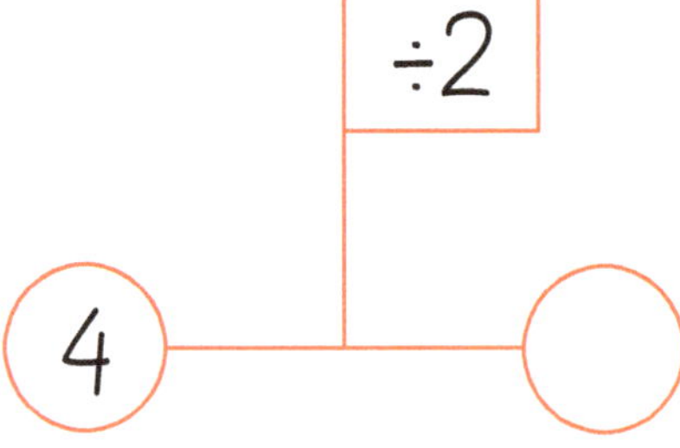

7.
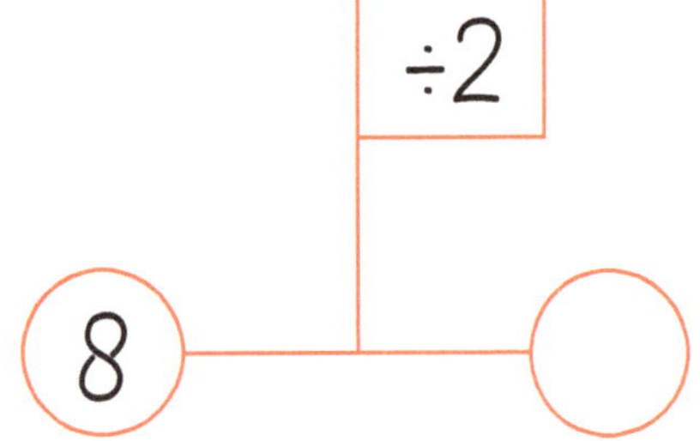

8.
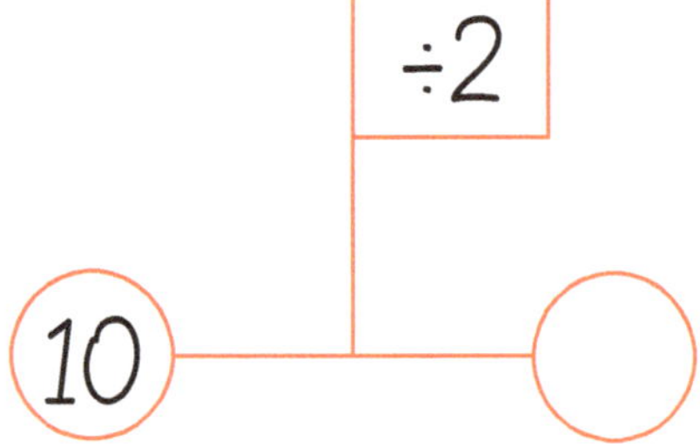

Flags
Level C

Find the output values below:

1. 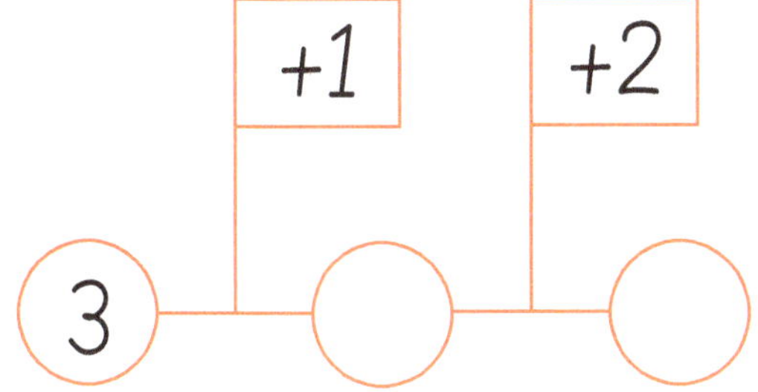

2. 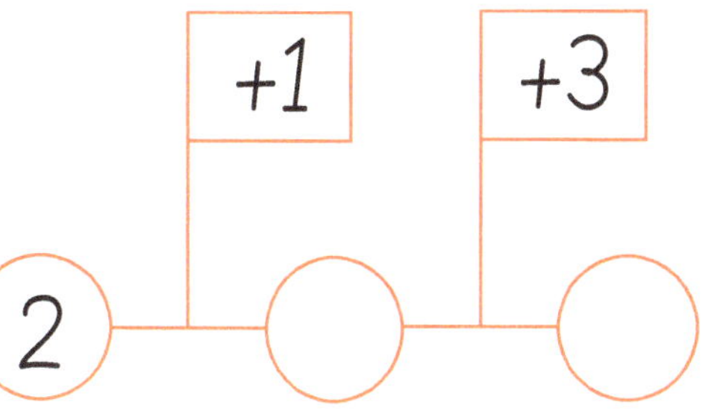

3. 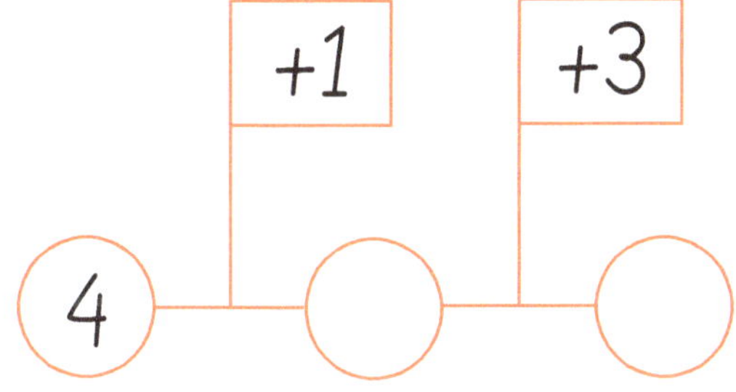

4. 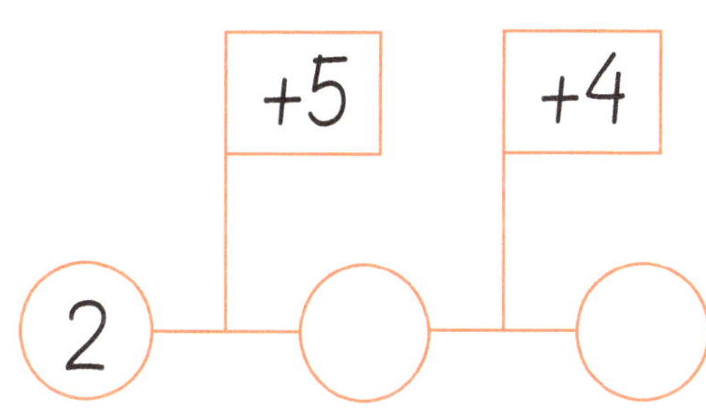

5. 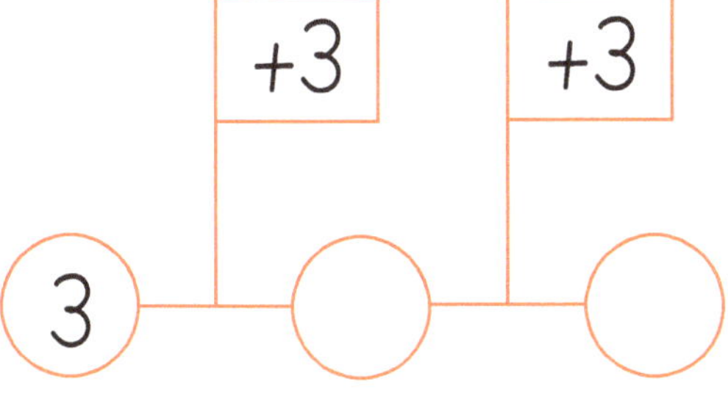

6. 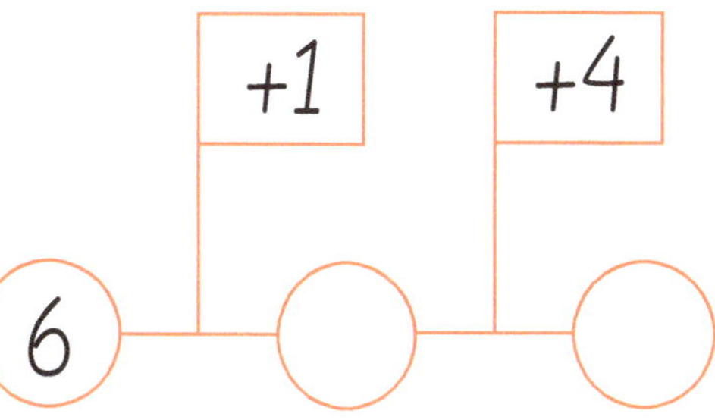

7. 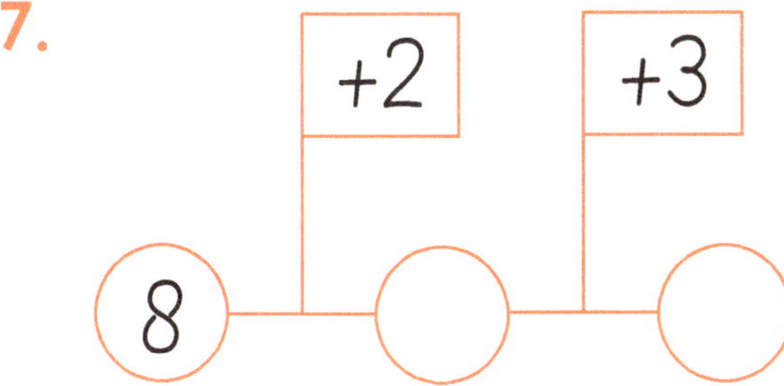

8. 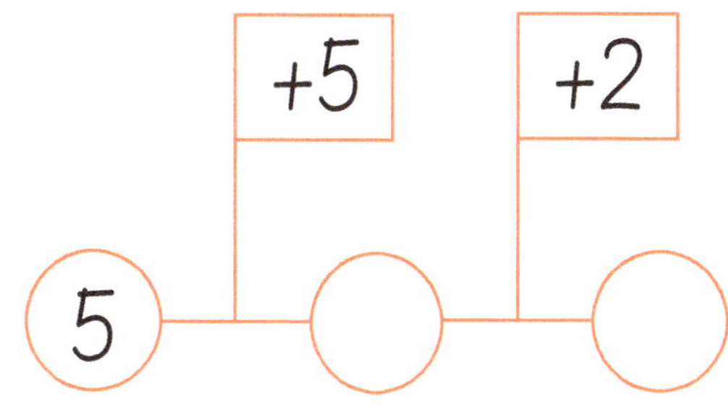

Flags
Level C

Find the output values below:

1. 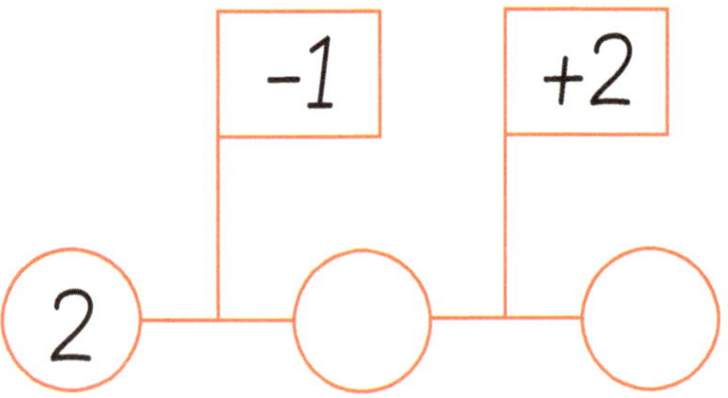

2. 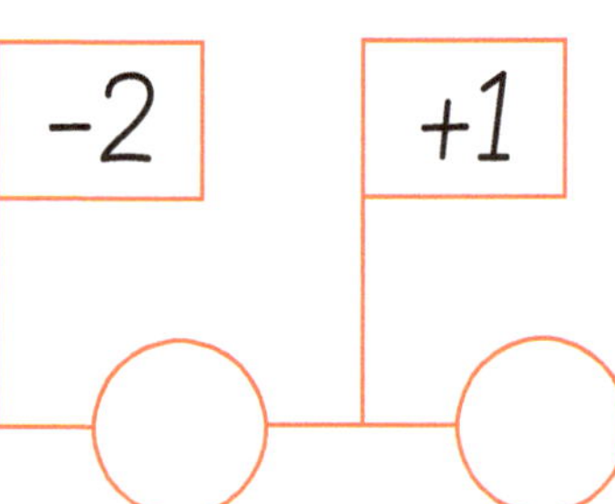

3. 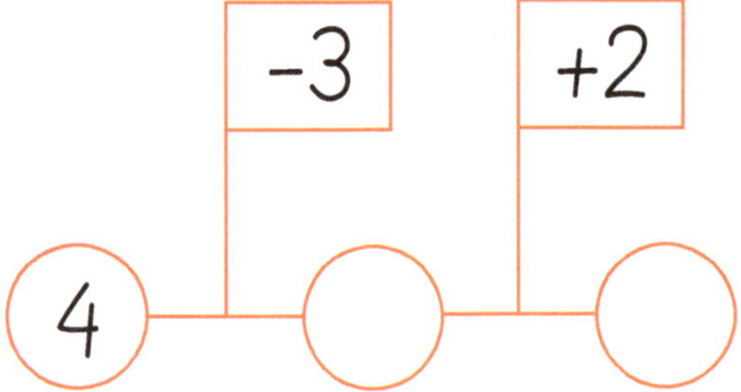

4. 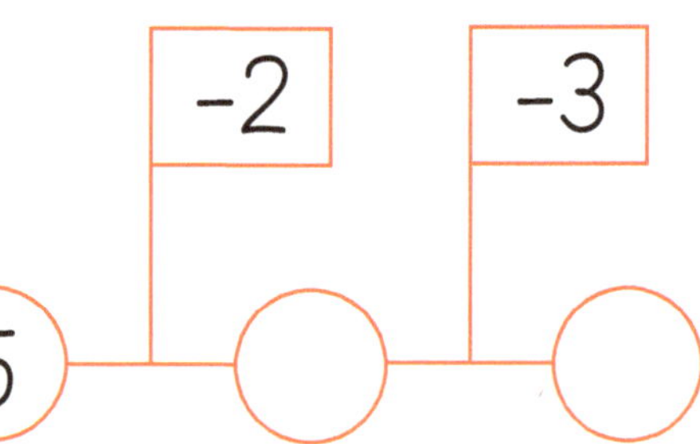

5. 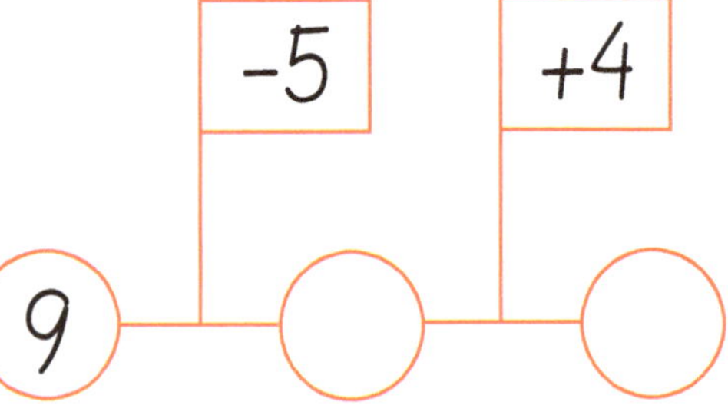

6. 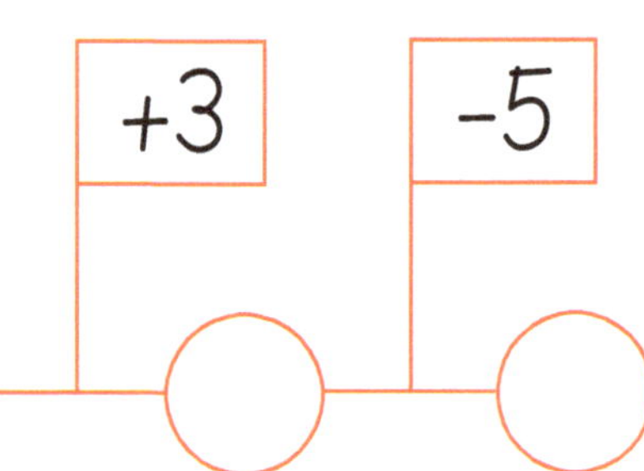

7. 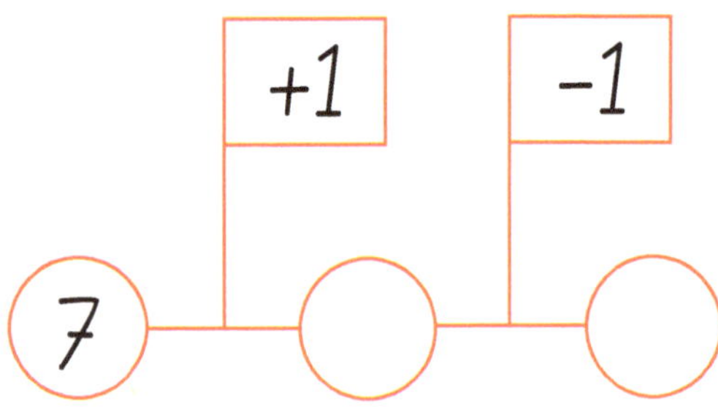

8. 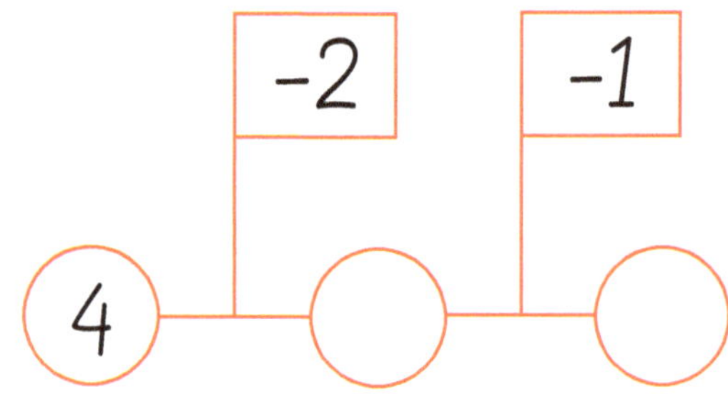

Flags
Level C

Find the output values below:

1. 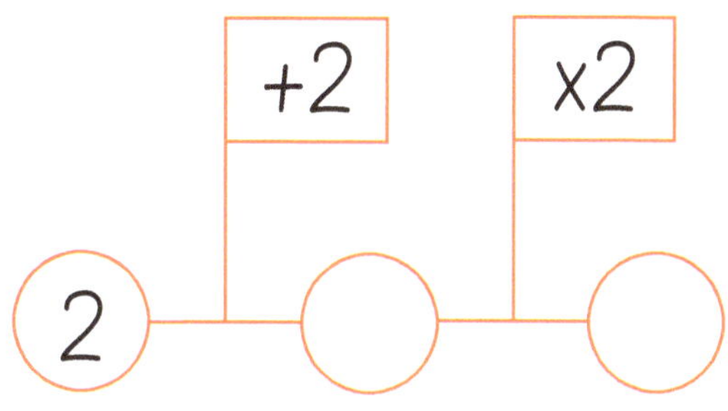

2. 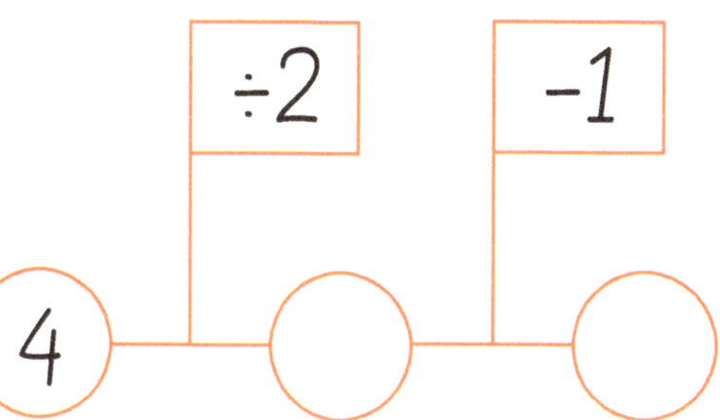

3. 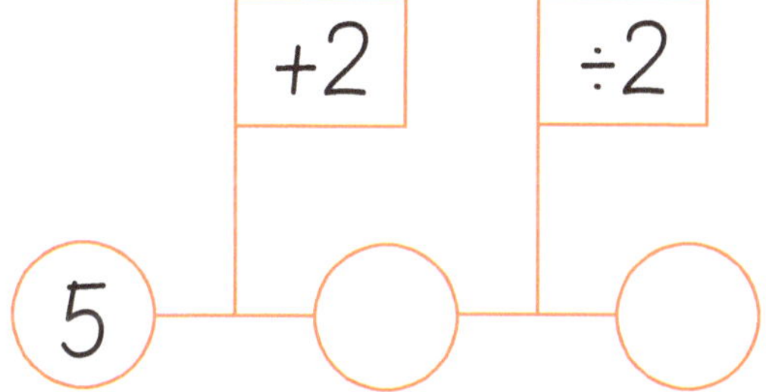

4. 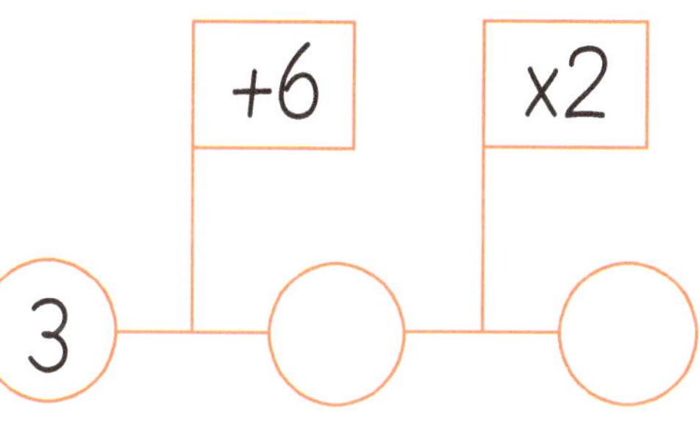

5. 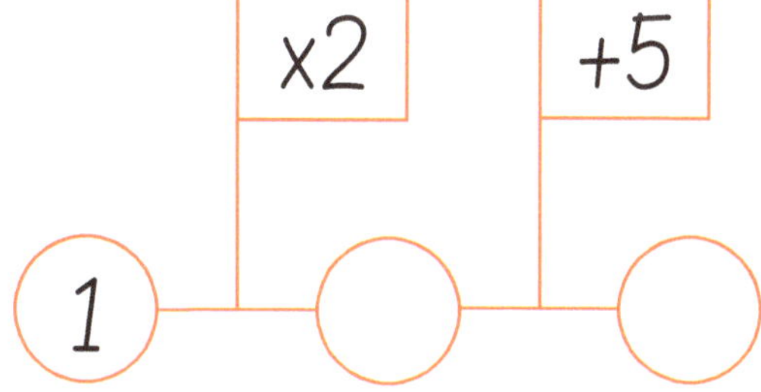

6. 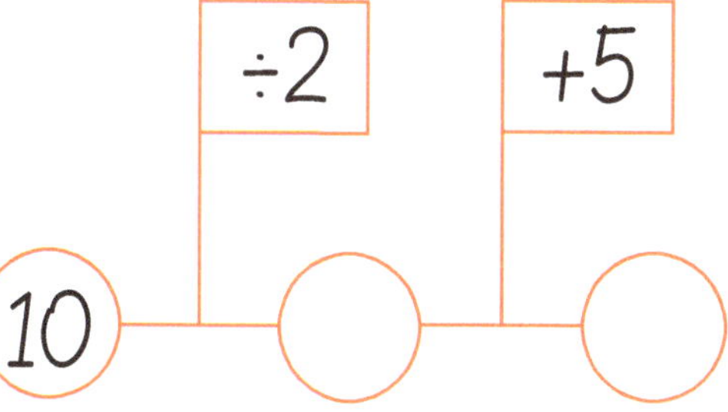

7. 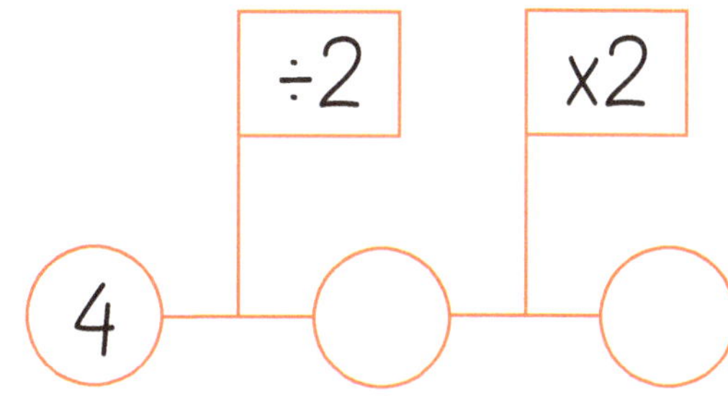

8. 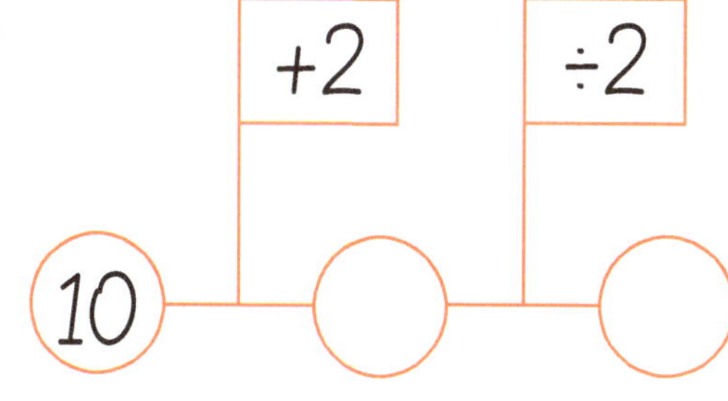

Flags
Level C

Find the output values below:

1.
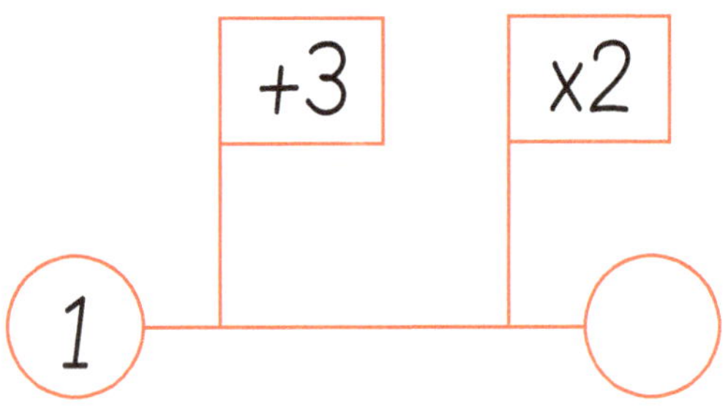

2.
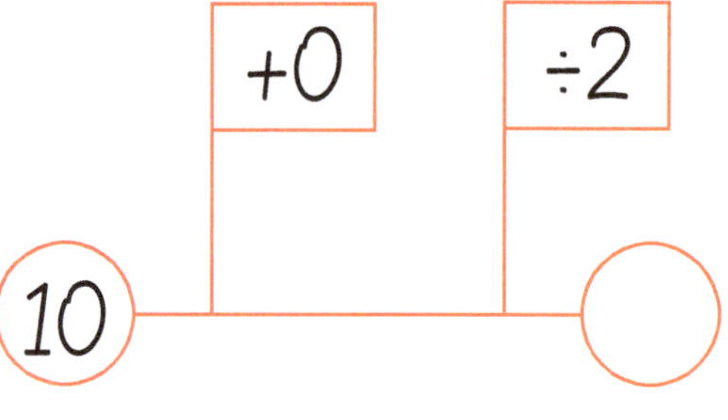

3.
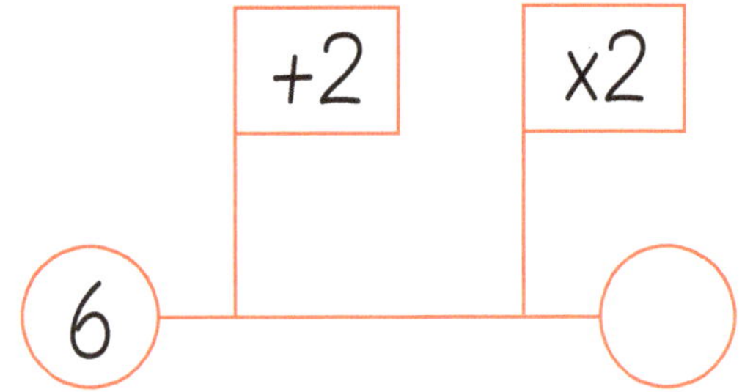

4.
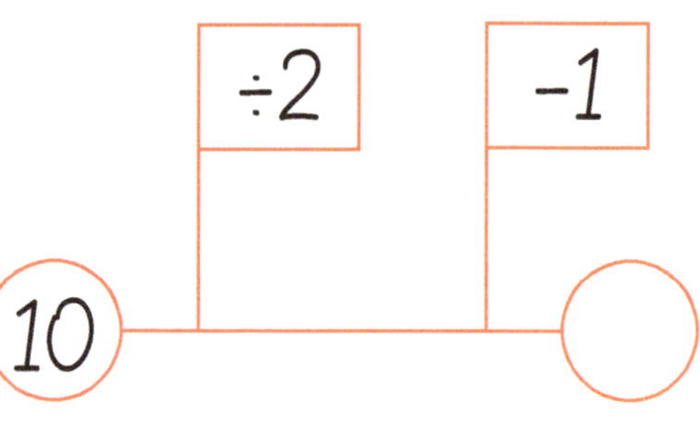

5.
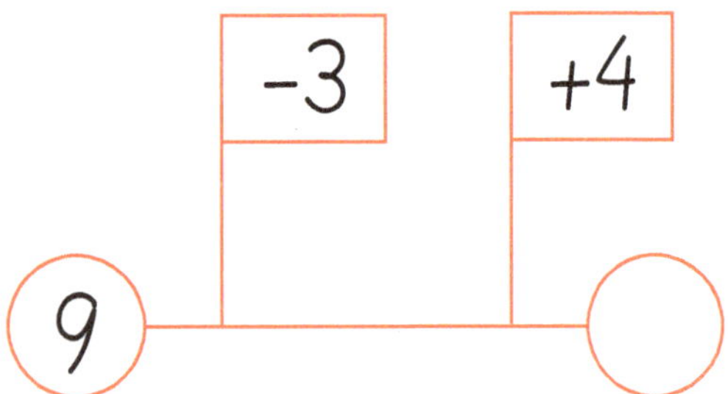

6.
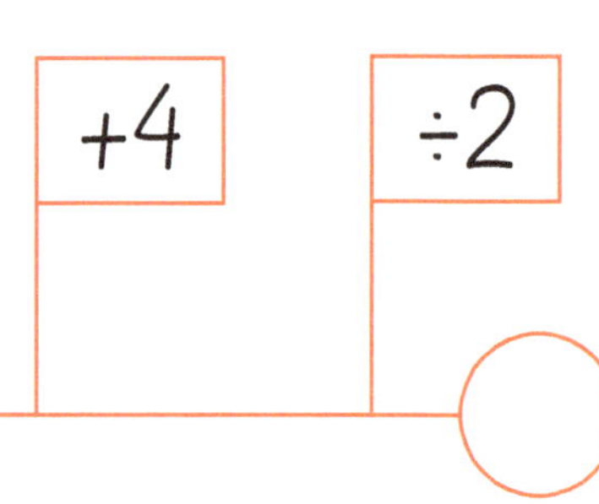

7.
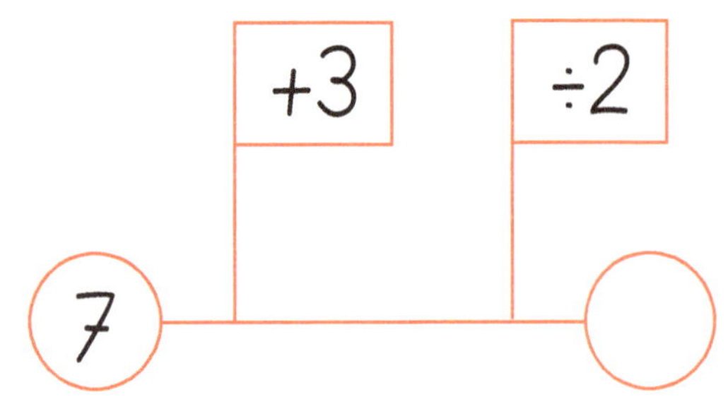

8.
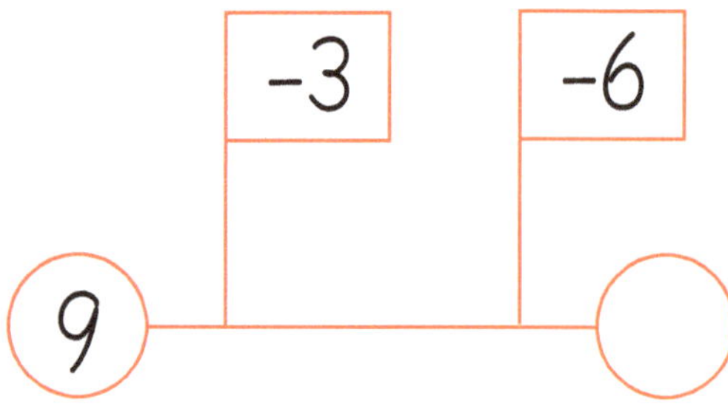

14

Flags
Level C

Find the output values below:

1. 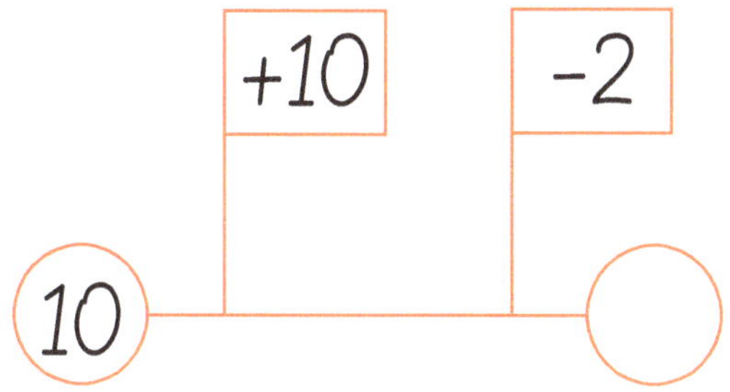

2. 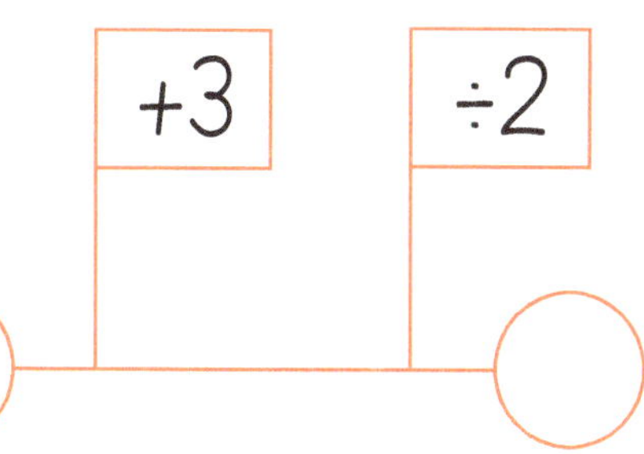

3. 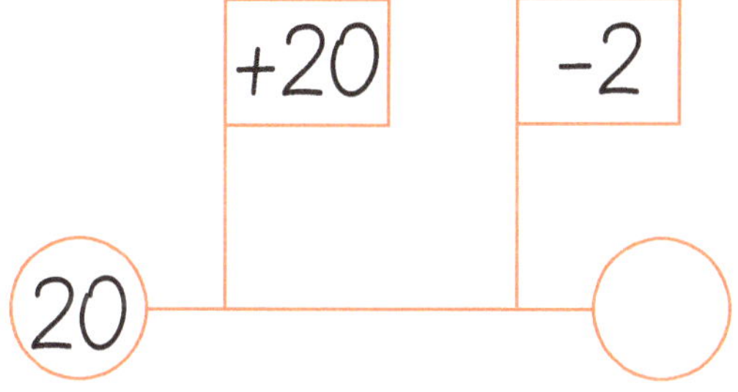

4. 

5. 

6. 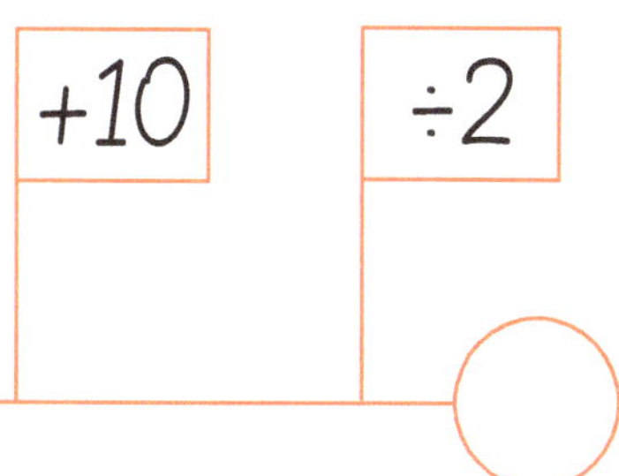

7. 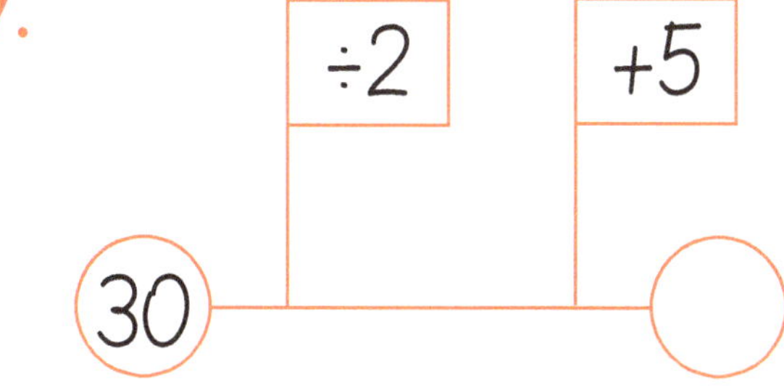

8. 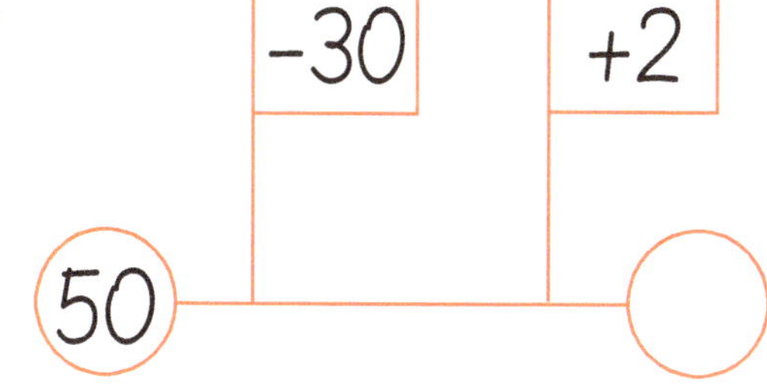

223

Flags
Level C

Find the output values below:

1.
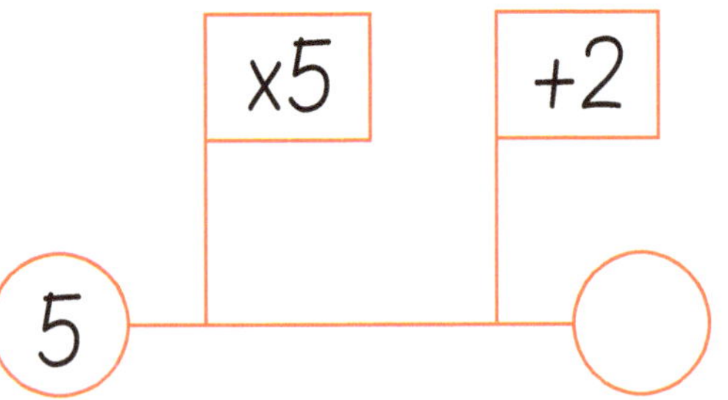

2.
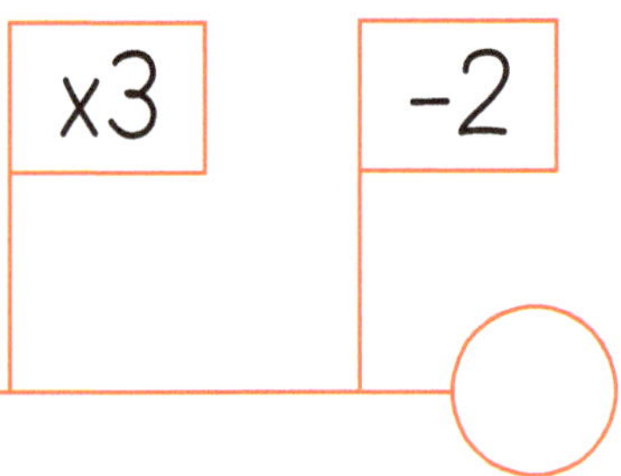

3.
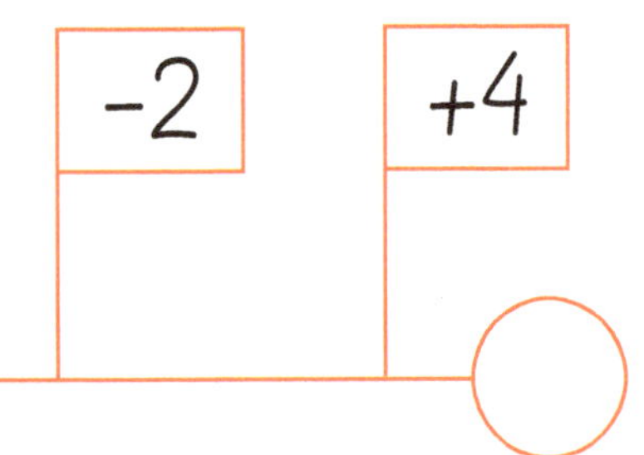

4.
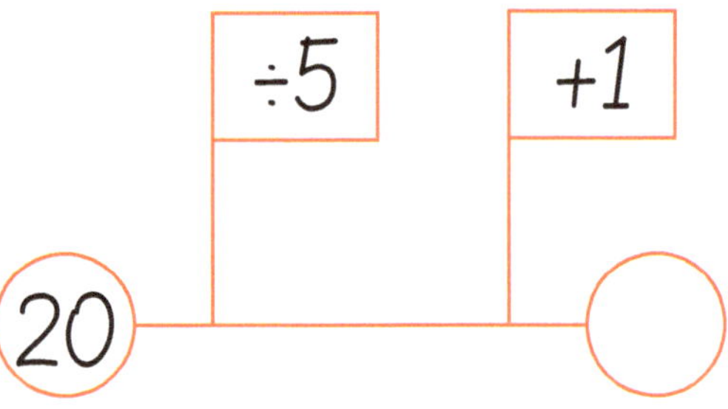

5.
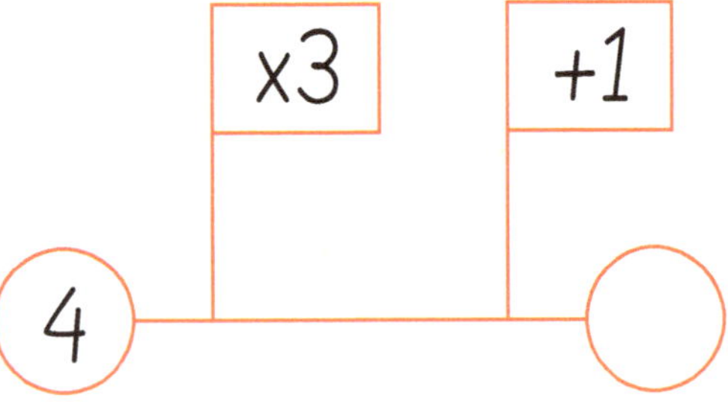

6.
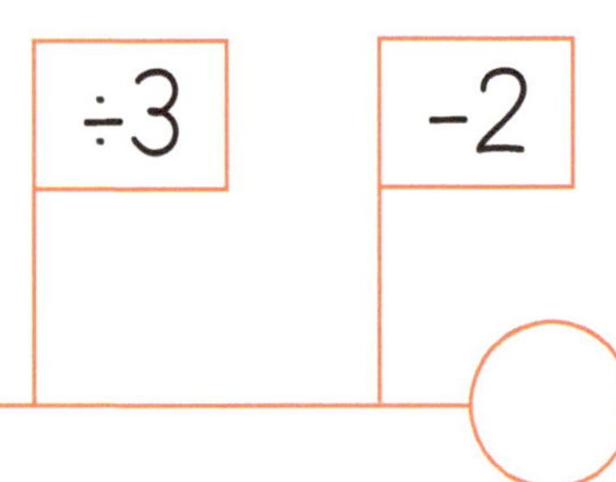

7.
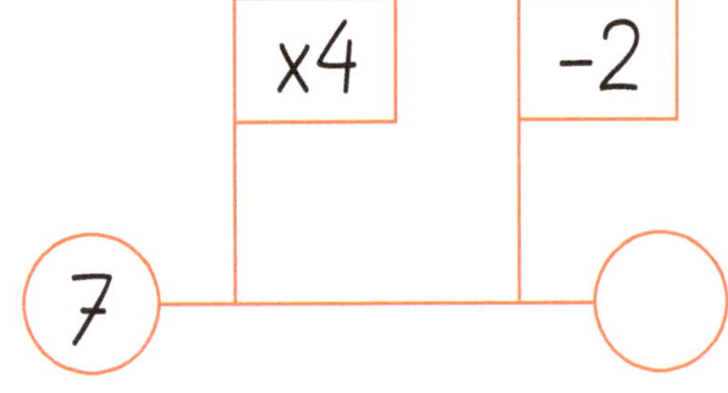

8.
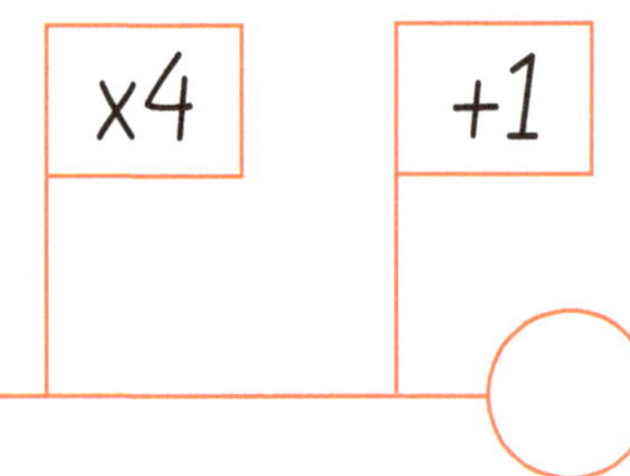

14

Problems

Ann has 5 apples. Her friend gave her a bag with more.
Altogether there were 8.
a. Draw a flag diagram to show the problem
b. How many apples were in the bag?

David has four sweets. He buys more in a packet.
Altogether there are ten sweets.
a. Draw a flag diagram to show the problem
b. How many sweets were in the packet?

Mark has three marbles. He finds a box with some more.
Altogether he has nine marbles.
a. Draw a flag diagram to show the problem
b. How many marbles were in the box?

Jane has 10p. She has more money in her purse. Altogether she has 15p.
a. Draw a flag diagram to show the problem
b. How much is in her purse?

Dawn has 2p. She has more money in her purse. Altogether she has 10p.
a. Draw a flag diagram to show the problem
b. How much money does she have in her purse?

Here are two processes.   +1,   x2

If the input number is 3, complete the flags to show how many output numbers you can find.

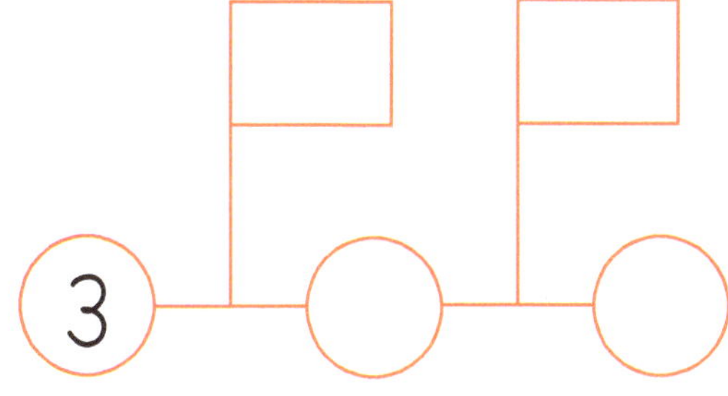

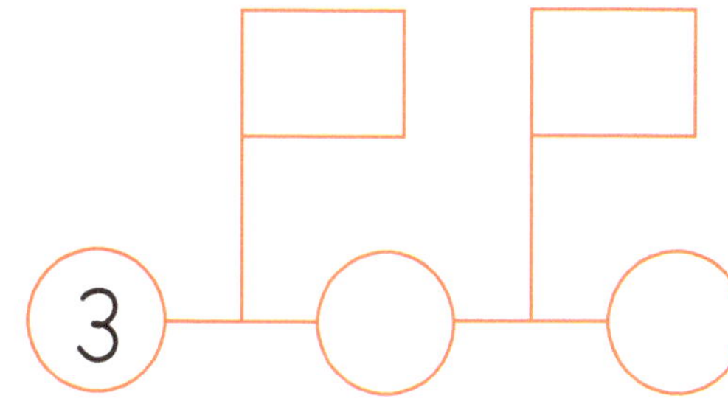

Problem

There were three cows in a field. The farmer let some more into the field. There where ten altogether.

a. Draw a flag diagram to show the problem
b. How many cows did the farmer let into the field?

Problems

Here are three processes. +1, +2, x2

If the input number is 5, complete the flags below to show how many output numbers you can find.

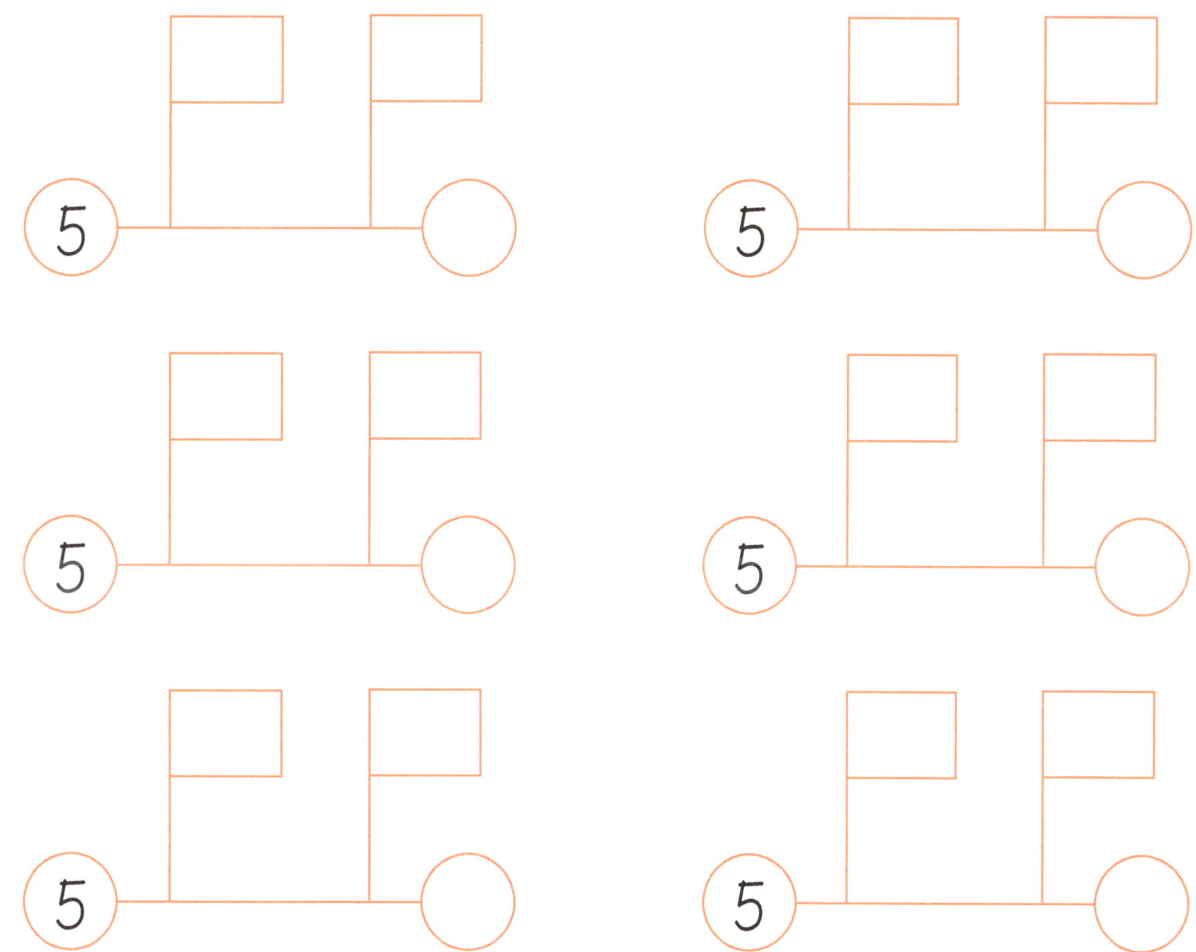

Here are two processes. ÷2, +4.

If the input number is 5, complete the flags below to show how many output numbers you can find.

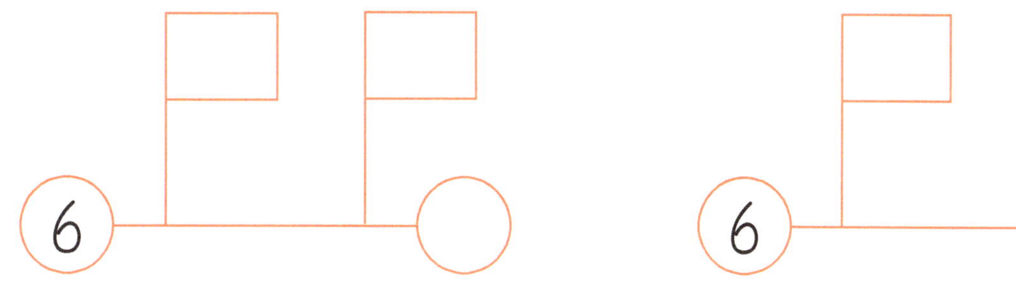

Problems

Find the output for the flag diagrams:

1.

2.

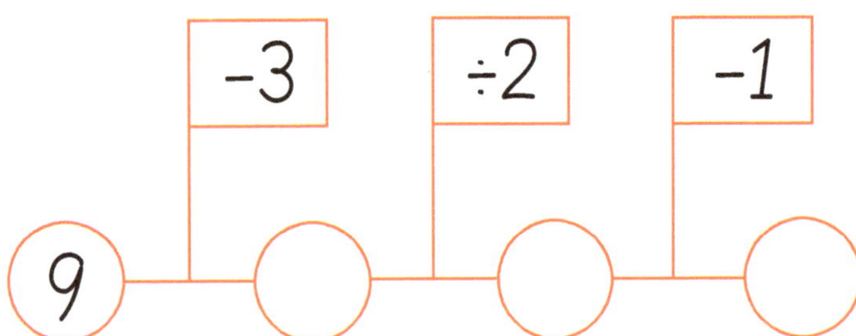

3.

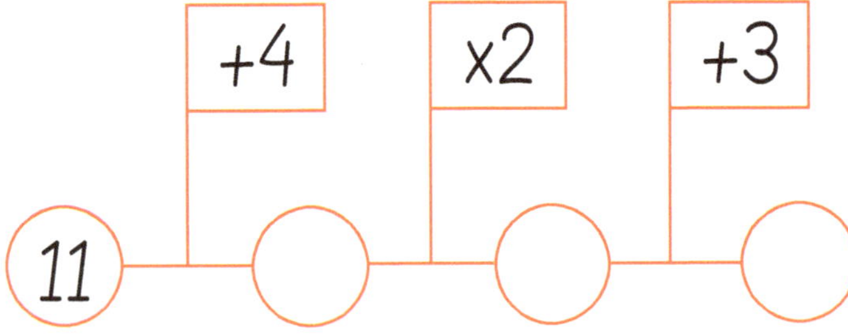

4.

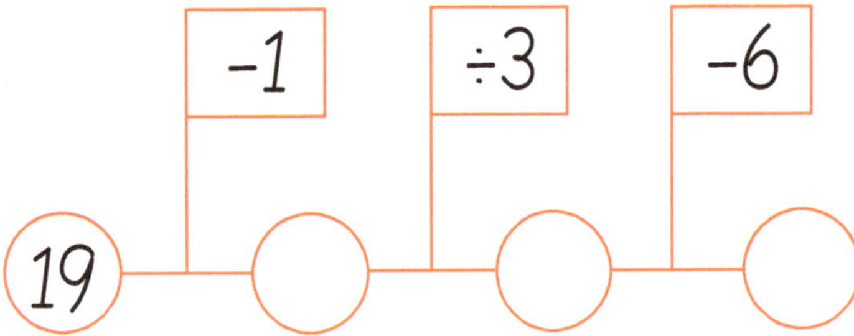

Problems

Find the output for the flag diagrams:

1.

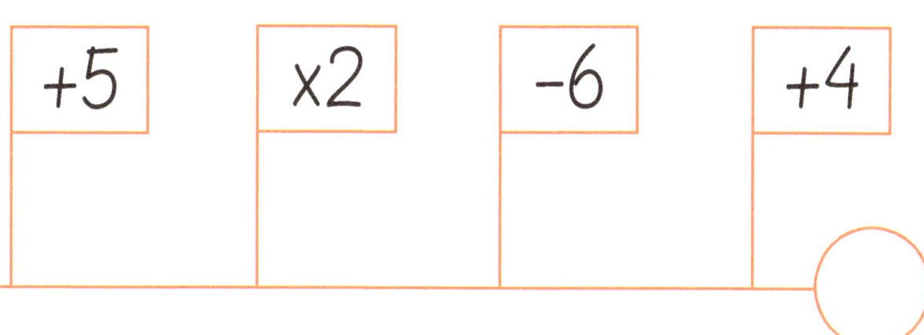

2.

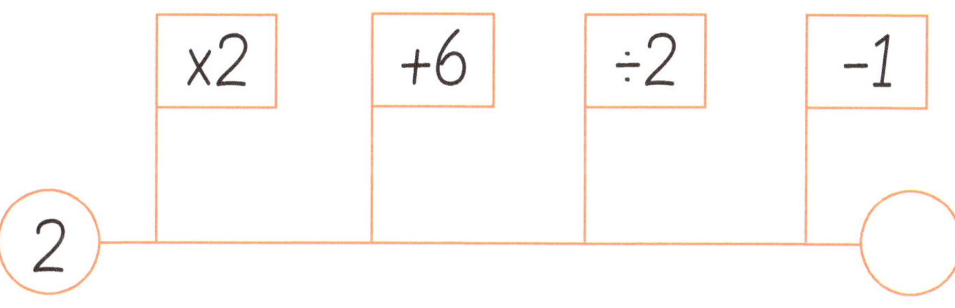

3.

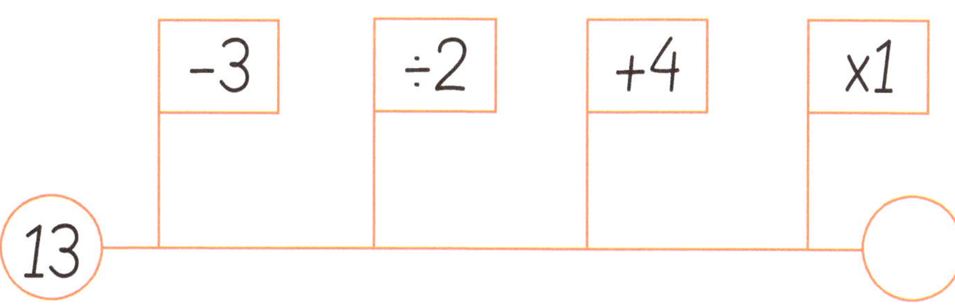

4.

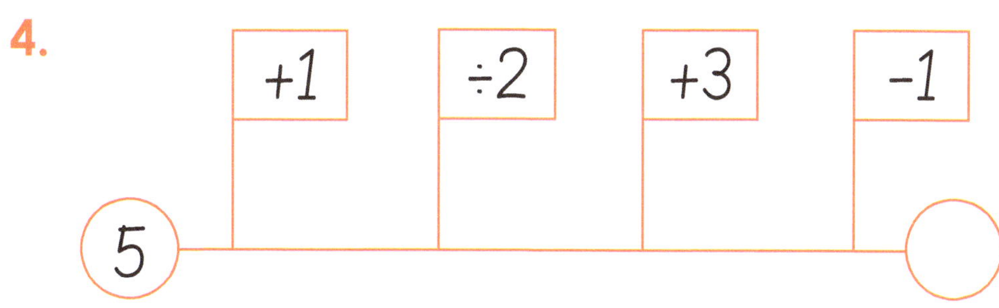

Problems

Here is an equilateral triangle of side 1cm

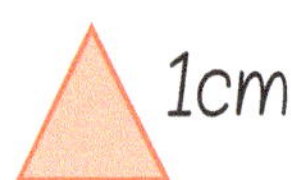

What is its perimeter? (HINT: Perimeter is the distance round the shape)

The perimeter of a triangle of side 1cm is ....

Here is an equilateral triangle of side 2cm

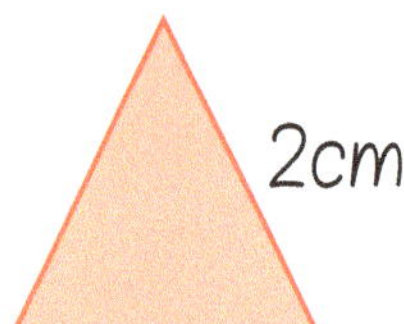

What is its perimeter? (HINT: Perimeter is the distance round the shape)

The perimeter of a triangle of side 2cm is ....

Here is an equilateral triangle of side 3cm

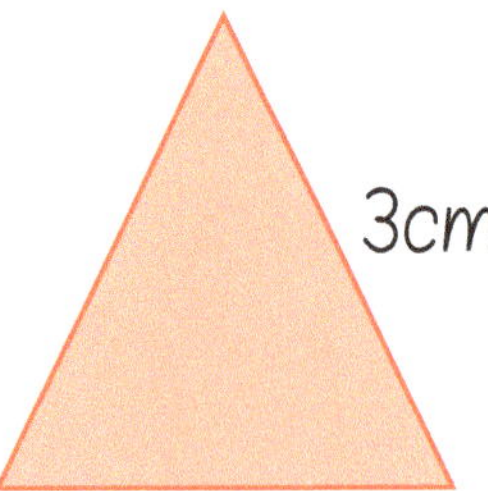

What is its perimeter? (HINT: Perimeter is the distance round the shape)

14

The perimeter of a triangle of side 3cm is ....

Results

The results can be recorded in a table as shown below:

Length of side (cm)	Perimeter of triangle (cm)
1	3
2	
3	
4	
5	
6	

Complete the table of results

Can you spot a pattern?

Complete the sentence below:

The perimeter of the triangle is .... times the............................

Extend the tables for triangles of side up to 10cm

Problems

Here is a square of side 1cm

1cm

What is its perimeter? (HINT: Perimeter is the distance round the shape)

The perimeter of a square of side 1cm is ....

Here is an square of side 2cm

2cm

What is its perimeter? (HINT: Perimeter is the distance round the shape)

The perimeter of a square of side 2cm is ....

Here is an square of side 3cm

3cm

What is its perimeter? (HINT: Perimeter is the distance round the shape)

The perimeter of a square of side 3cm is ....

Results

The results can be recorded in a table as shown below:

Length of side (cm)	Perimeter of square (cm)
1	4
2	
3	
4	
5	
6	

Complete the table of results

Can you spot a pattern?

Complete the sentence below:

The perimeter of the SQUARE is .... times the ....

Extend the tables for square of side up to 10cm

Notes

Notes

235

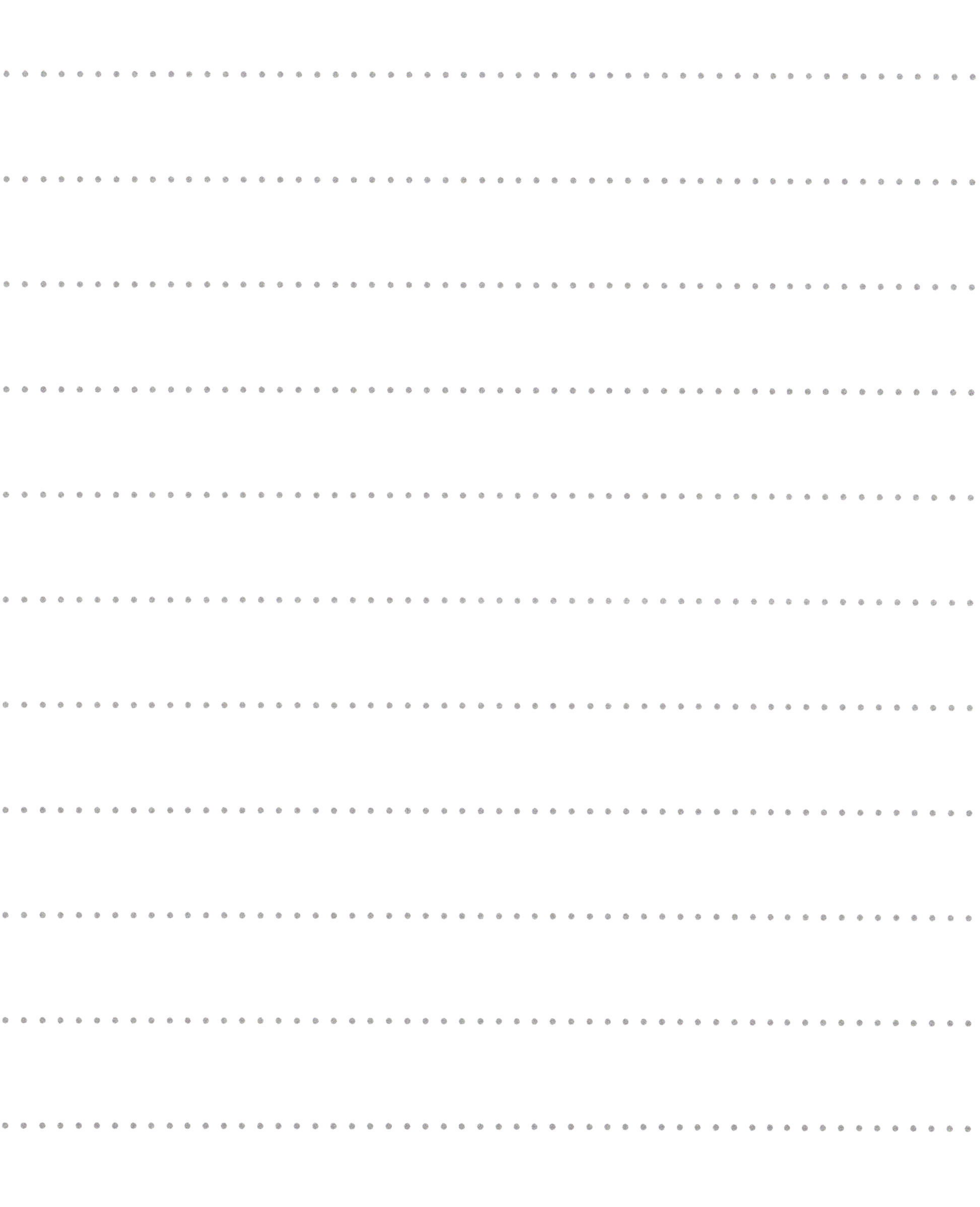

Unit 15

Functions and Equations

Teacher Note

Functions and equations

Outcome: Number, money and measurement

Strand: Functions and equations

Target: Recognise and explain simple relationships between two sets of numbers or objects.

Aims

So that the pupil can understand reverse operations i.e. that addition is the inverse operation to subtraction and that division is the inverse operation to multiplication.

Prior Knowledge

The pupil should know how a number machine operates and how to find the OUTPUT using a flag diagram given the INPUT and PROCESS.

Introduction

An explanation of what number machine is should be given with 2-3 examples. The relation between a number machine and a flag diagram should be shown and explained.

Read the following examples carefully

Find the input for the flag diagrams below:

1.
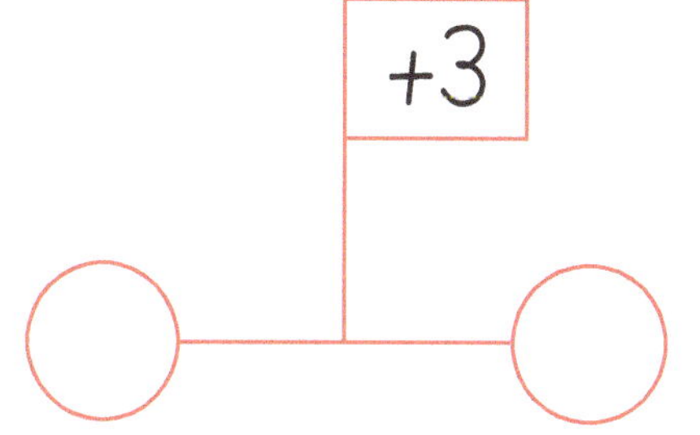

The input is 7 because 7 + 3 = 10

2.
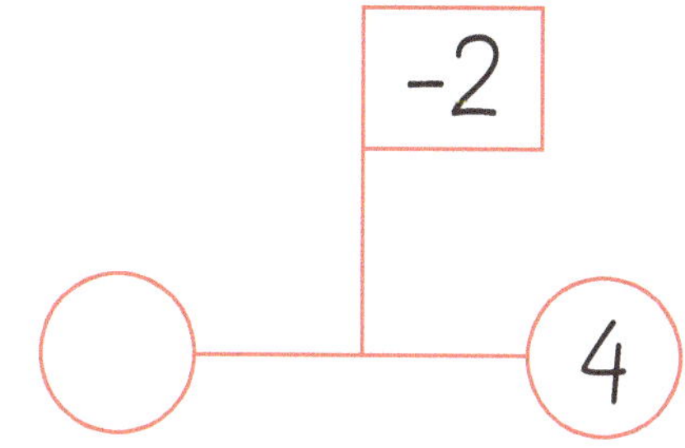

The input is 6 because 6 - 2 = 4

Find the input for the flag diagrames below:

1.
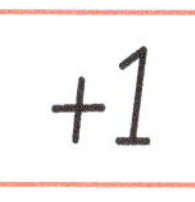

2.

3.

4.
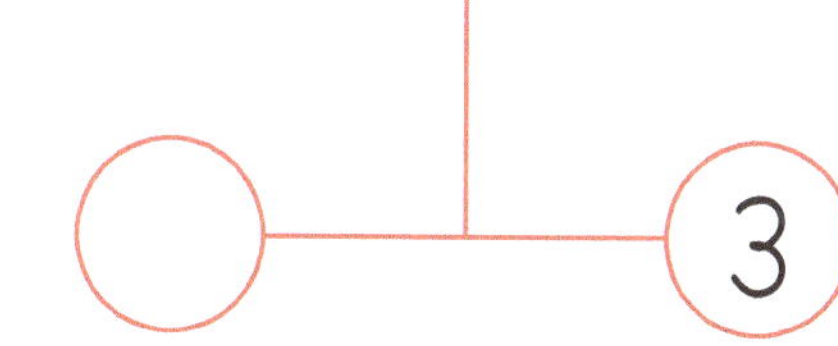

5.
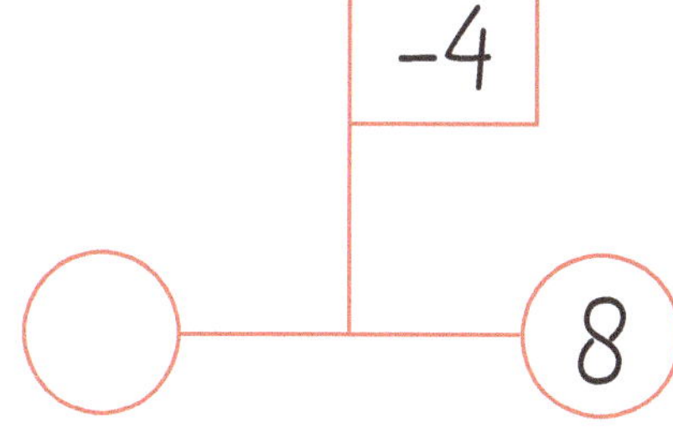

6.
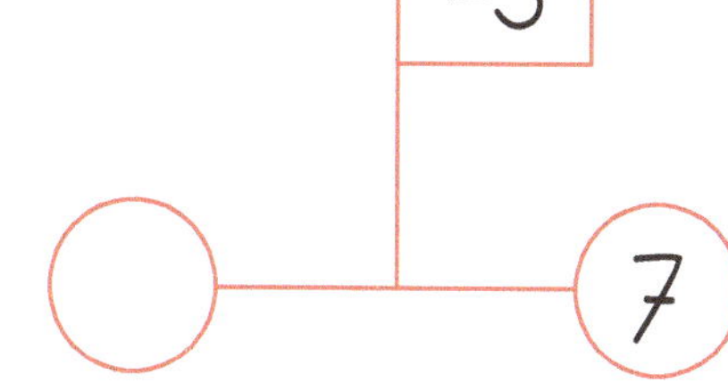

7.
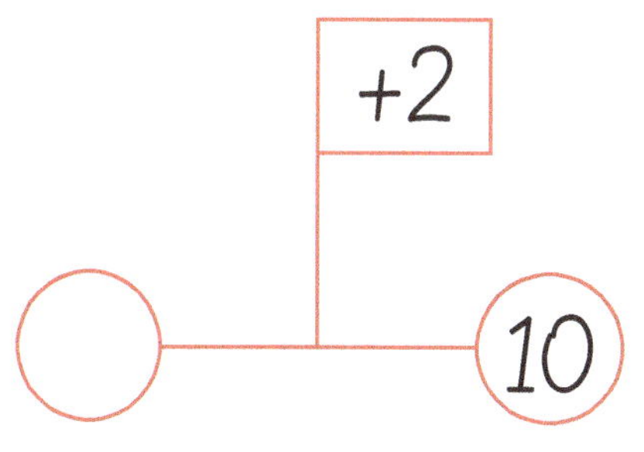

8.
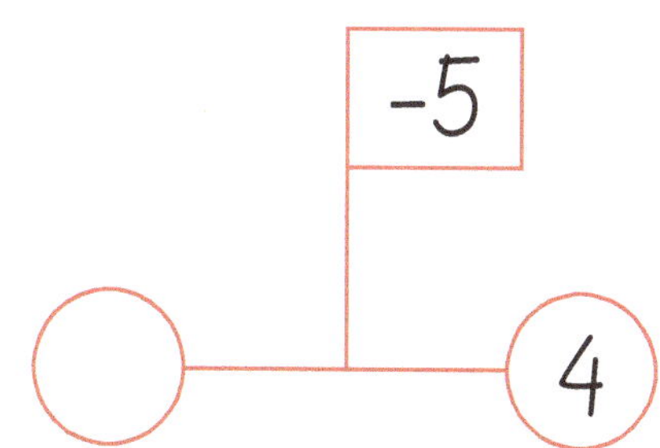

15

Flags
Read this page carefully

1. Find the output for each of the flag diagrams below:

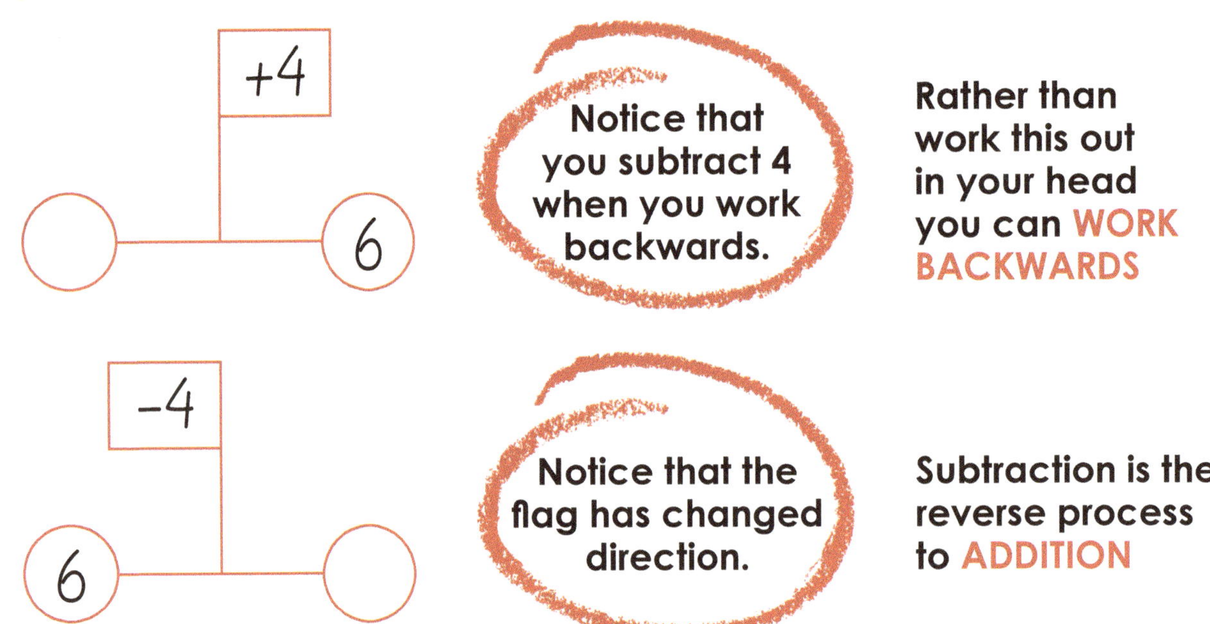

---

2. Calculate the input for this flag diagram:

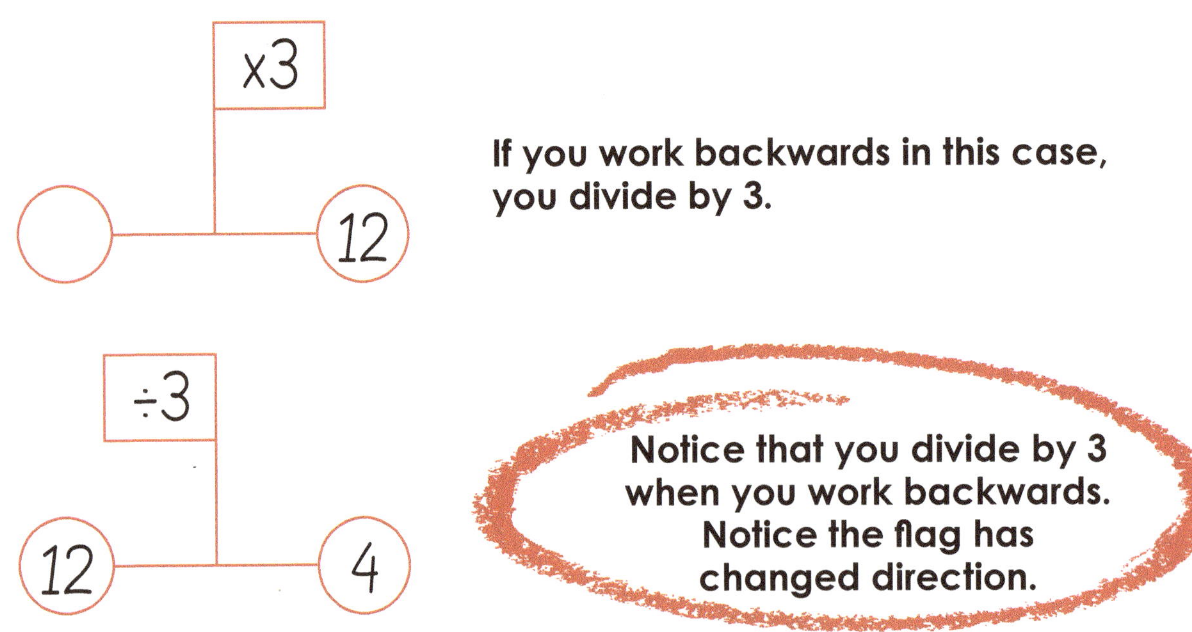

Division is the reverse process to Multiplication.

Flags
Level D

By working backwards in each case, calculate the input.
The first flag diagram has been started for you.

1.

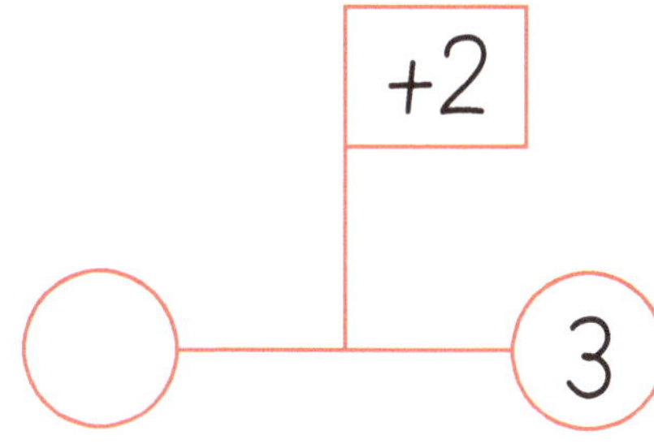

2.

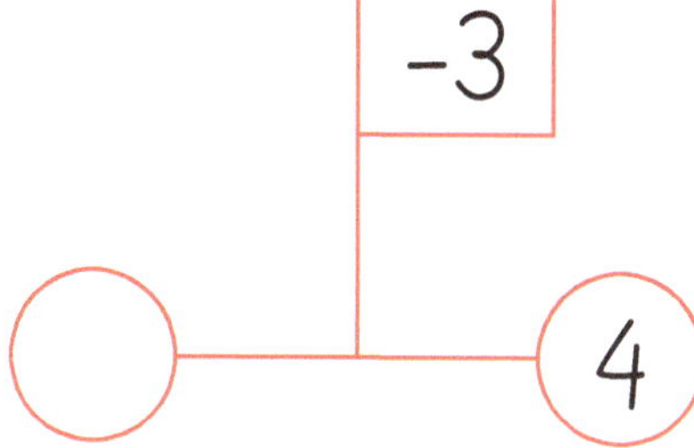

3.

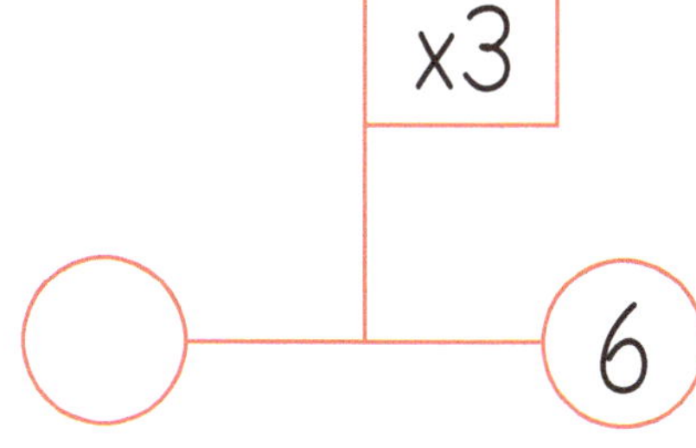

4.

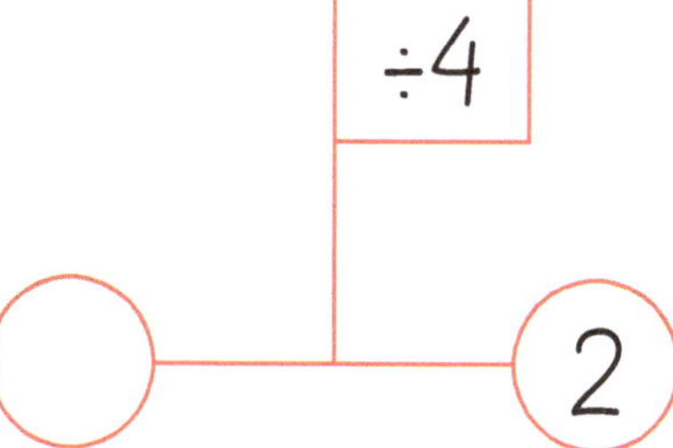

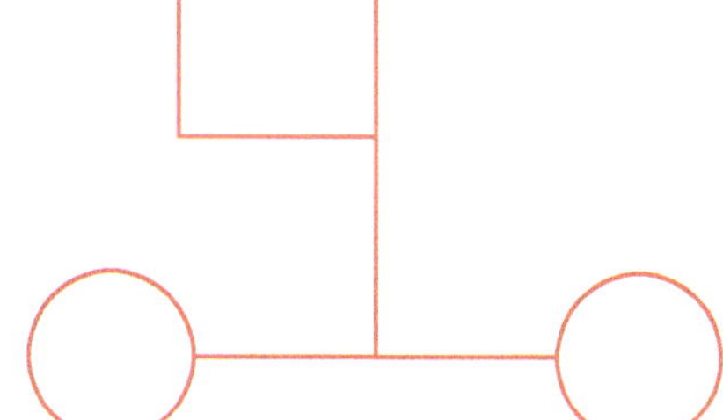

15

Flags
Level D

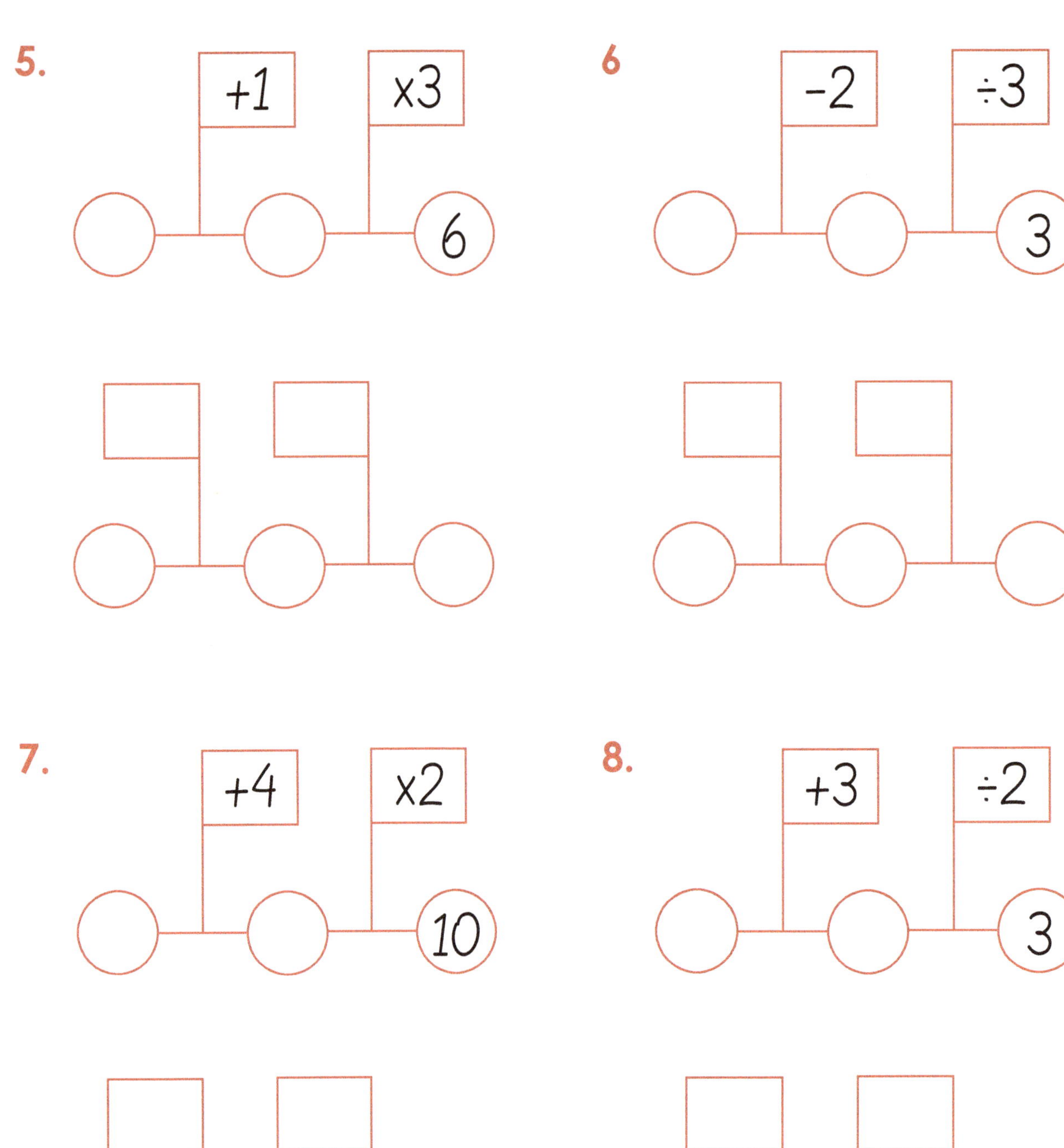

Flags
Read this page carefully

Find the input for the flag disgrams below?

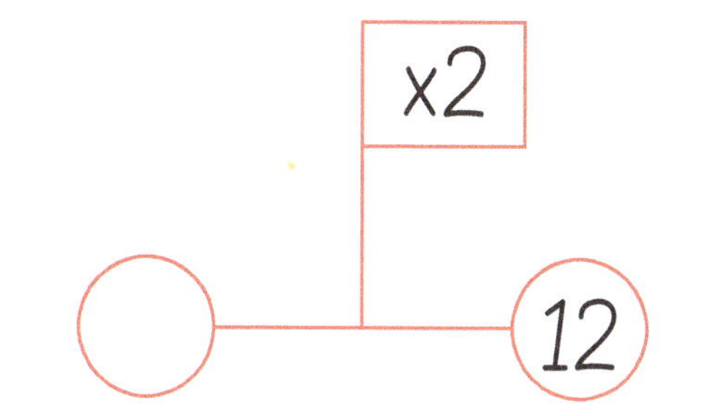

The input is 6 because 6 x 2 = 12

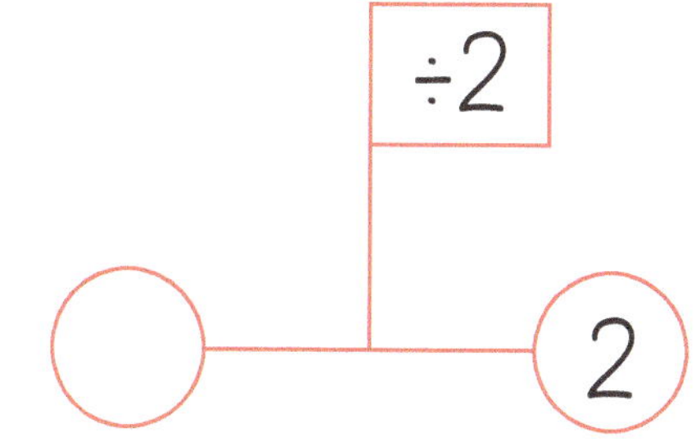

The input is 8 because 8 ÷ 4 = 2

Find the input for the flag disgrams below?

1.

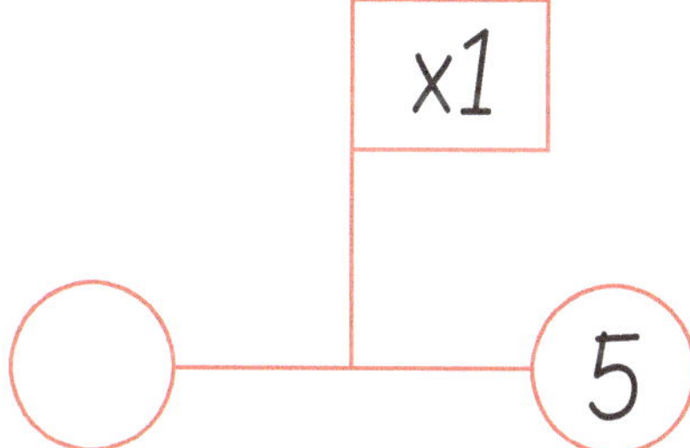

2.

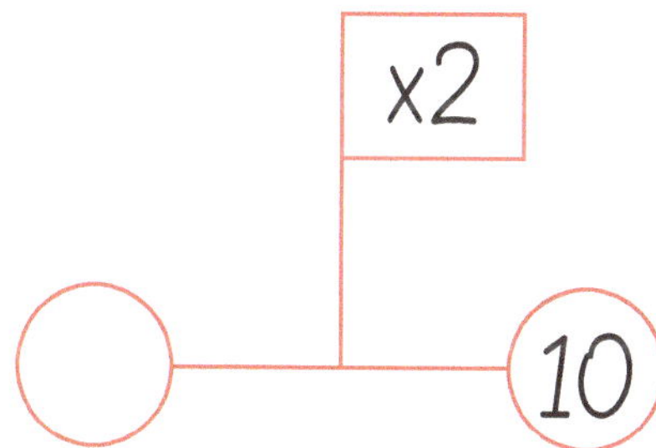

3.

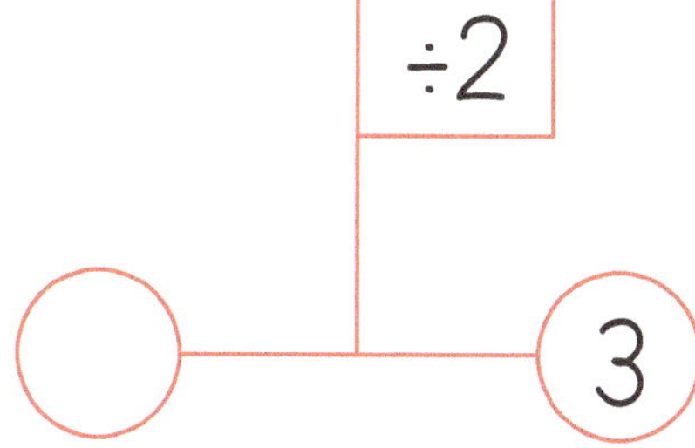

4.

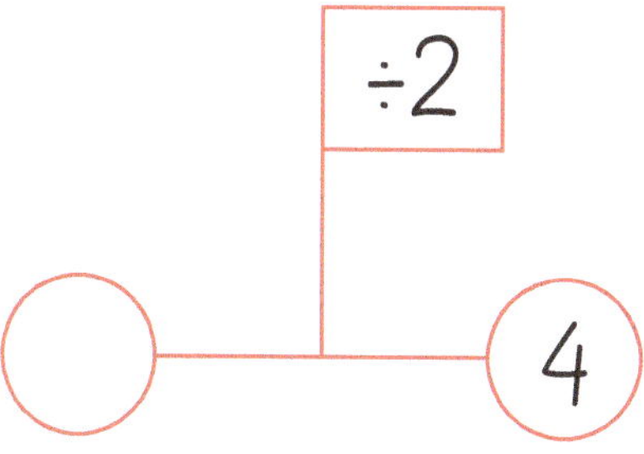

5.

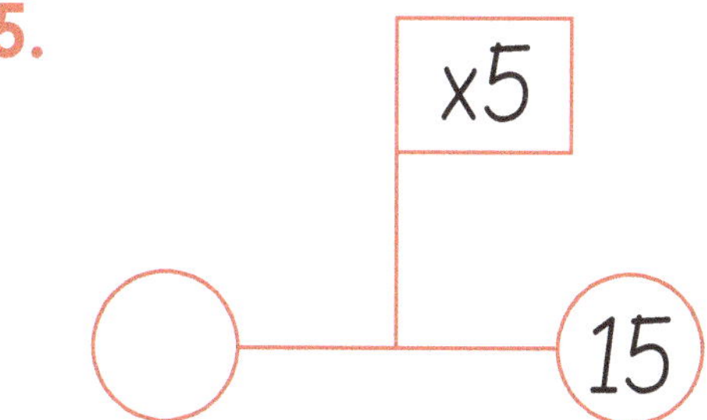

6.

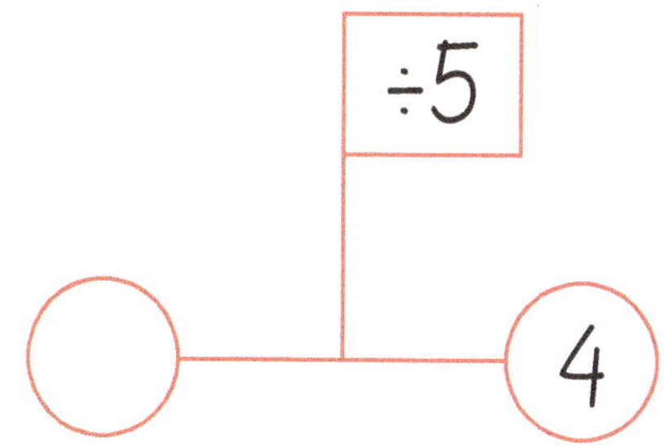

15

Flags
Level D

Find the input for each flag diagram below by working backwards:

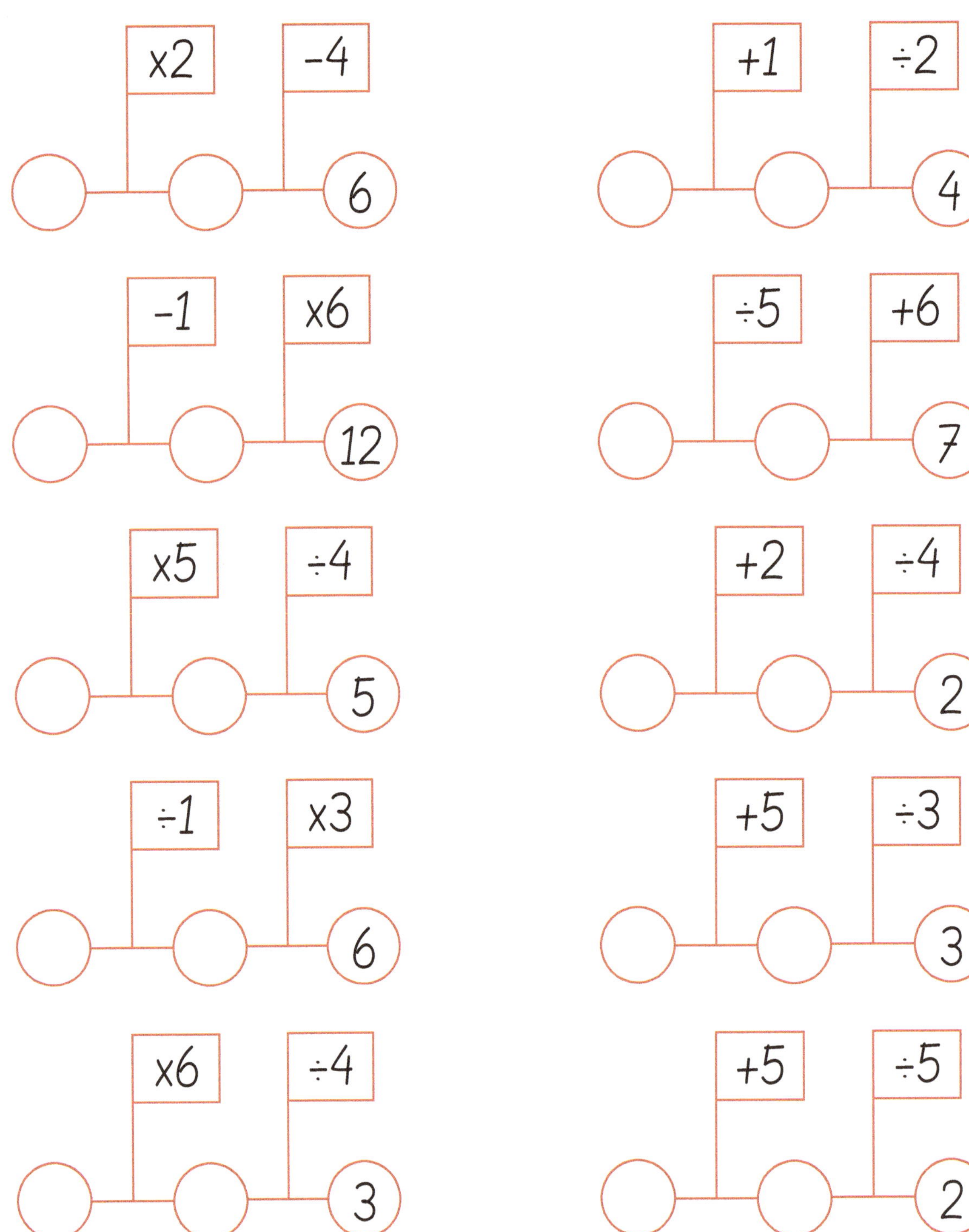

Flags
Level D

Find the input for each flag diagram below by working backwards:

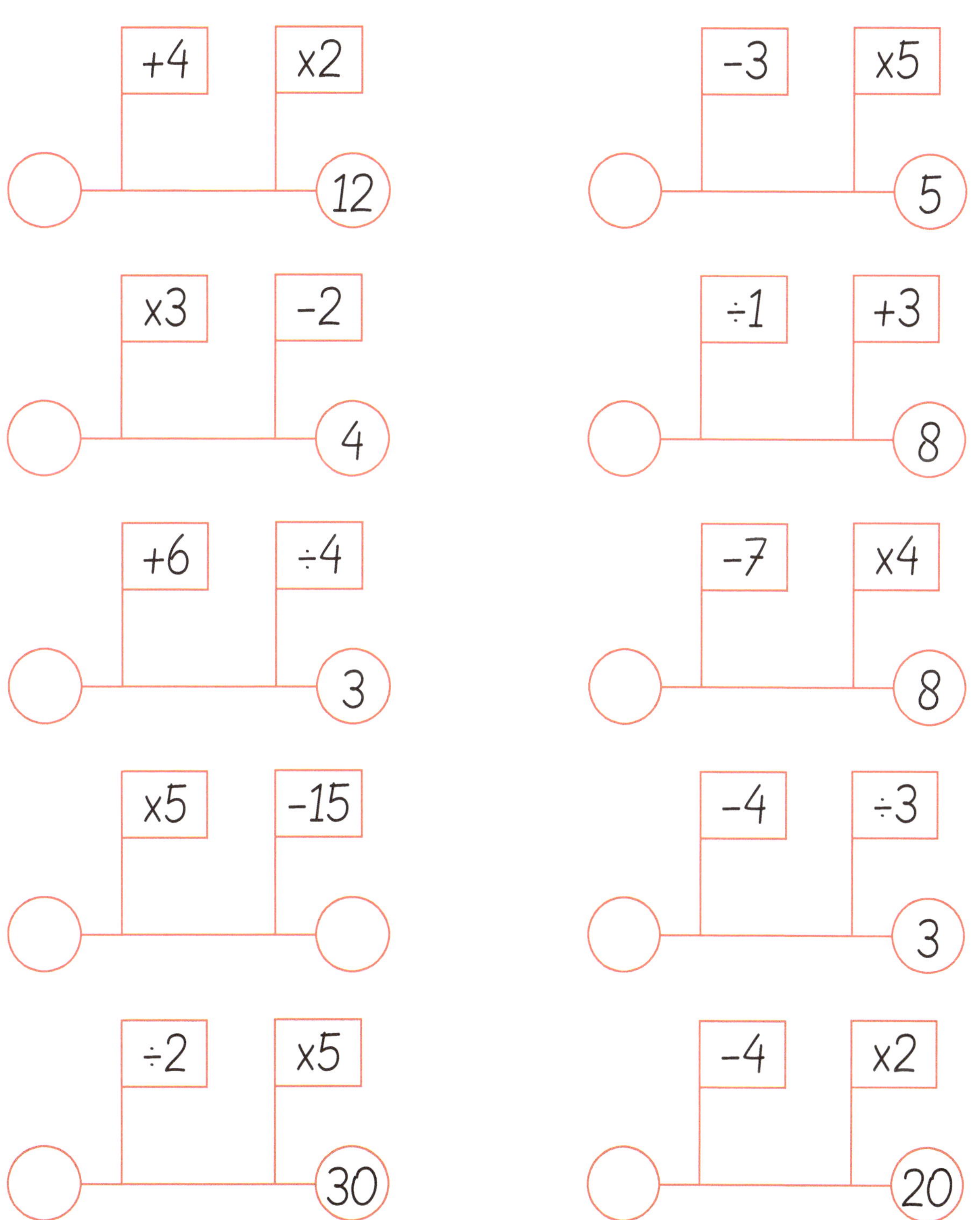

15

Flags
Level D

Here are three number machines:

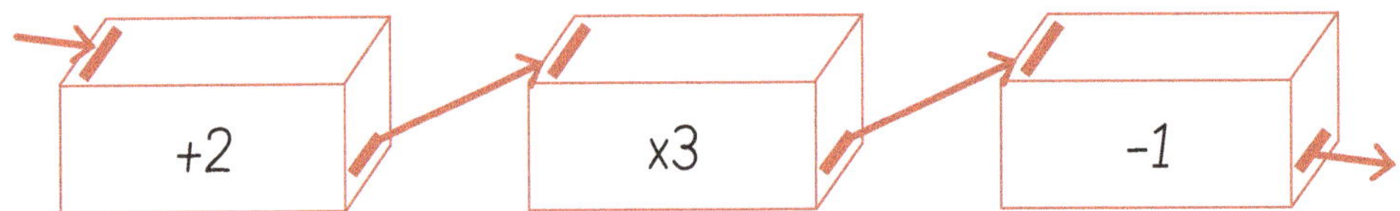

These calculations can be shown as a flag diagram.

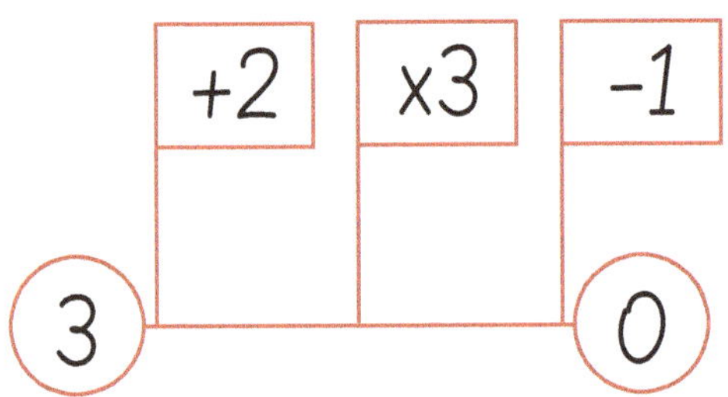

What is the output? Write it in the output circle above.

The three number machines can be used in different orders. There are six different ways. Show them by completing the flag diagrams below.

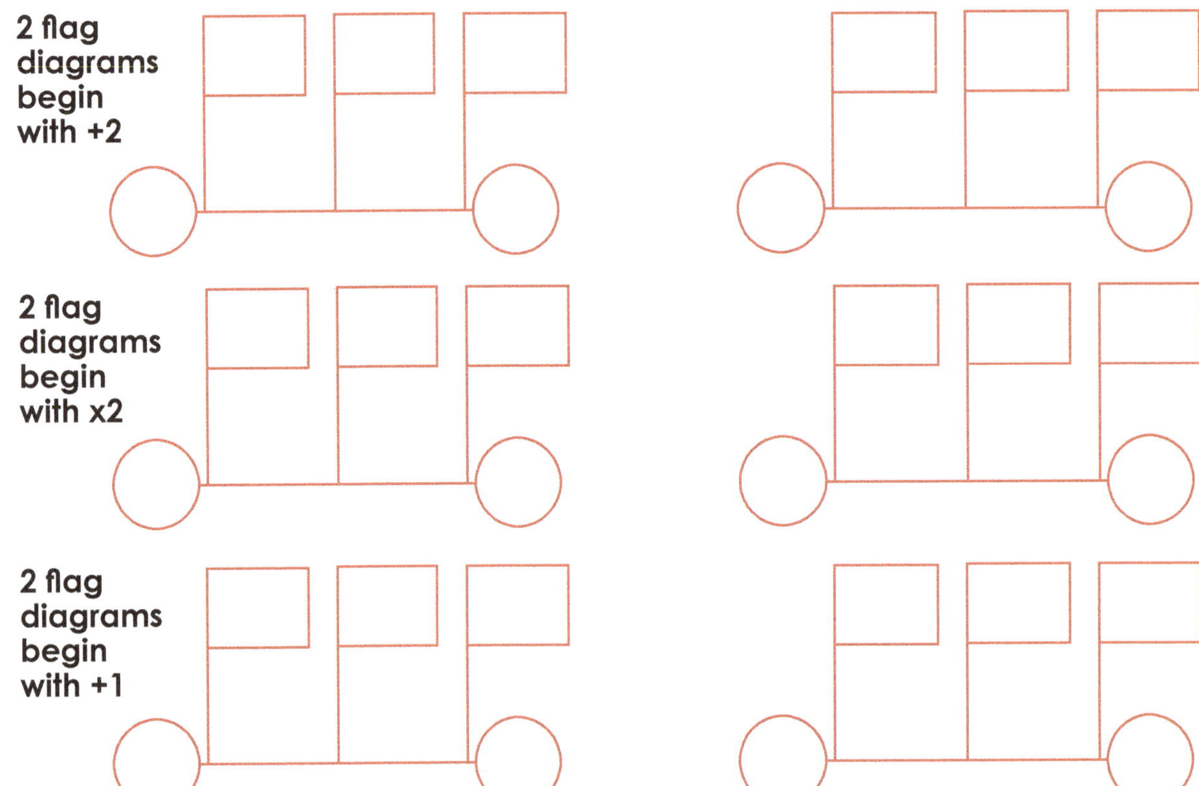

Flags
Level D

Compare your answers with the flag diagrams below:

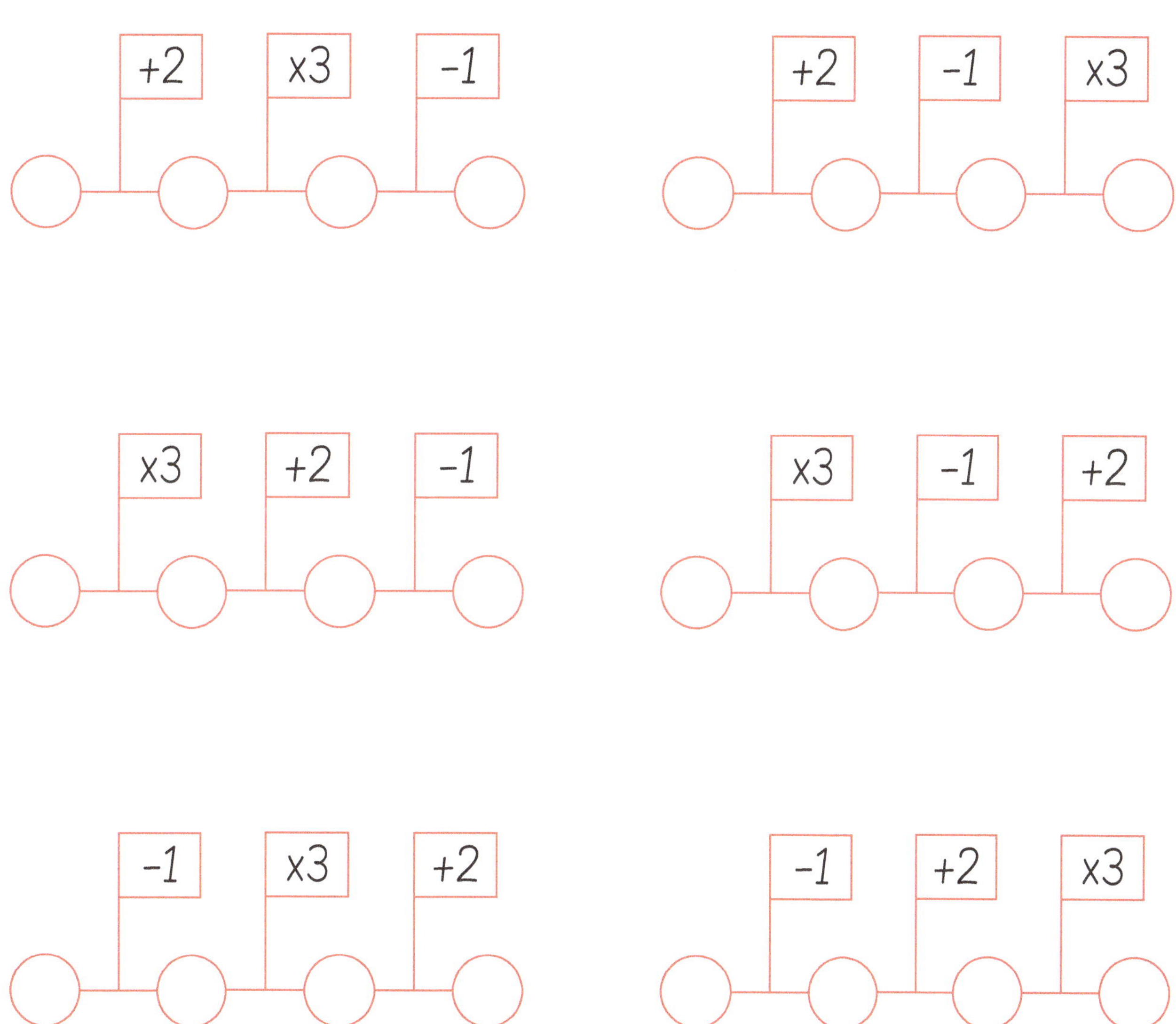

Calculate the output for every flag diagram above.

Flags
Level D

You have three flags available.

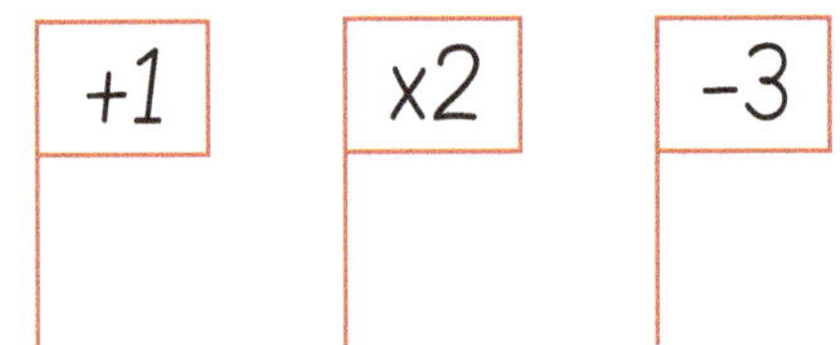

a. Complete all six flag diagrams below using the three flags.
b. Calculate the output for each flag diagram.

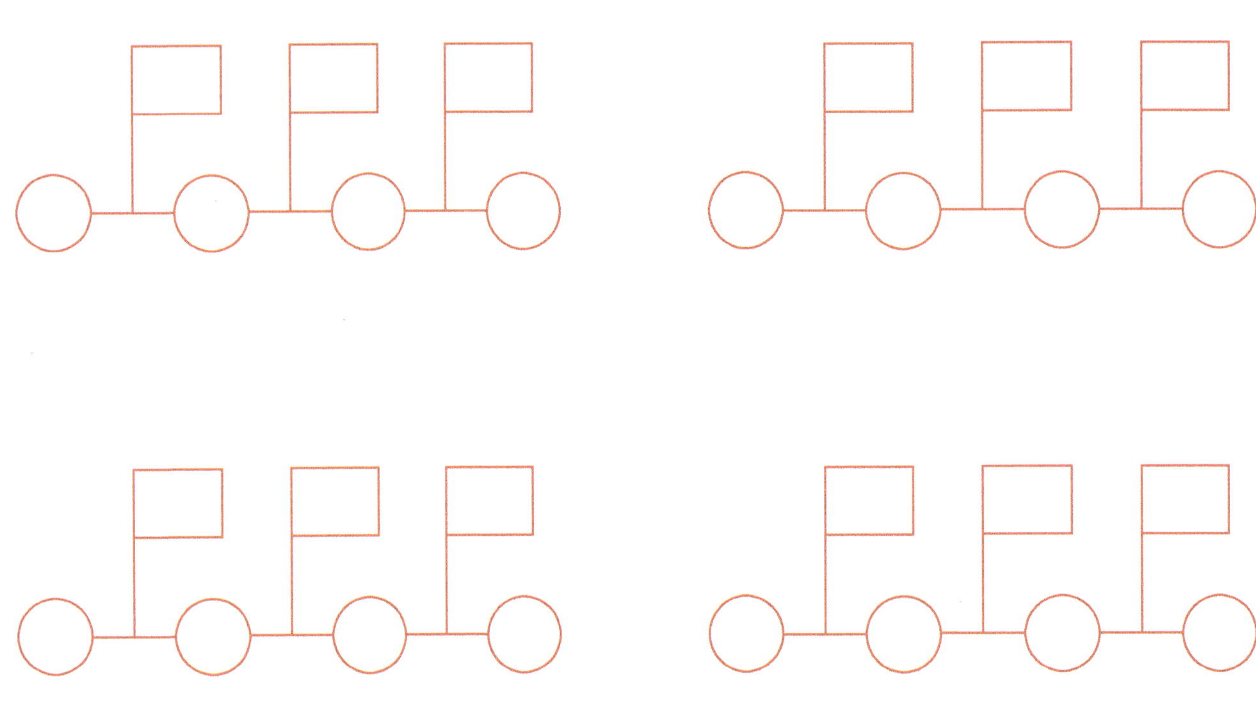

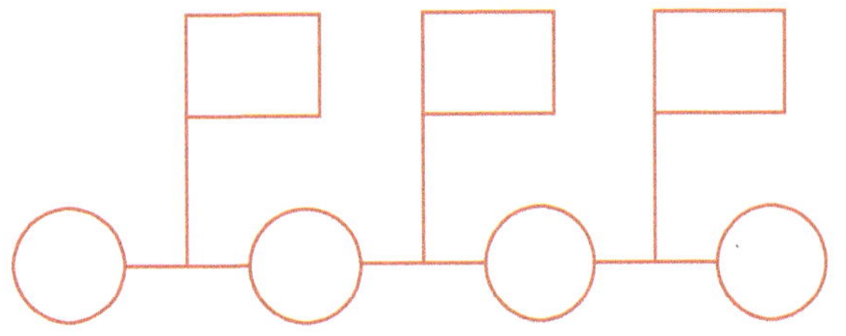

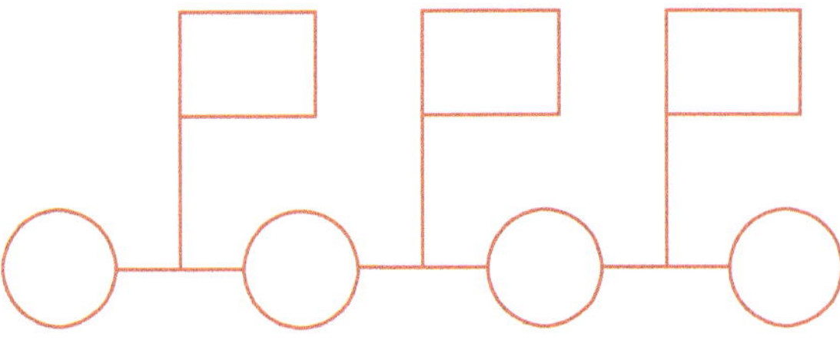

Flags
Level D

You have three flags available.

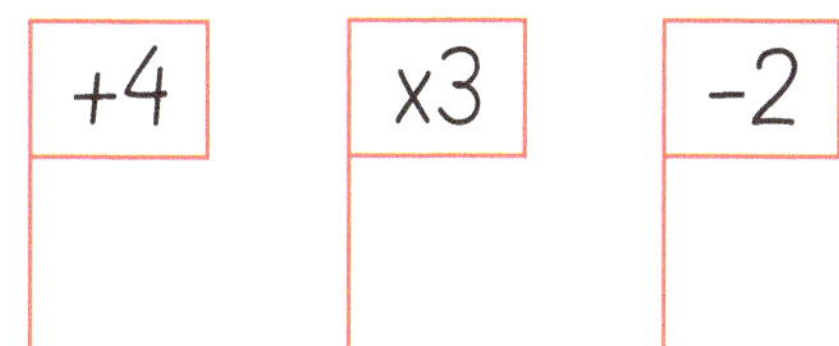

a. Complete all six flag diagrams below using the three flags.
b. Calculate the output for each flag diagram.

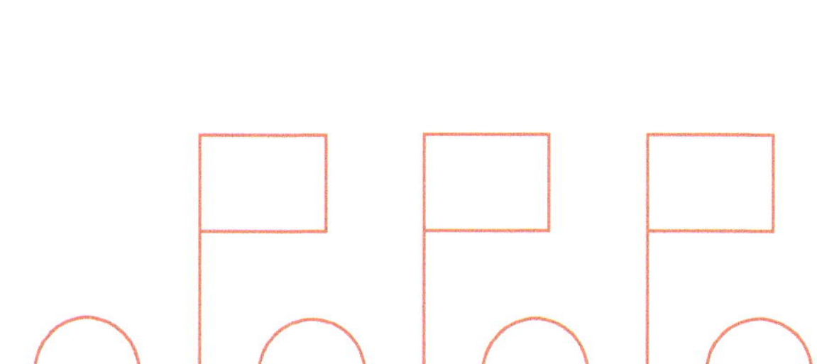

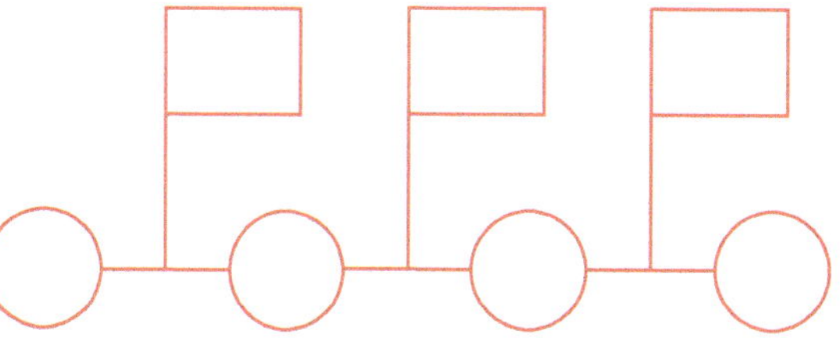

15

Flags
Level D

You have three flags available.

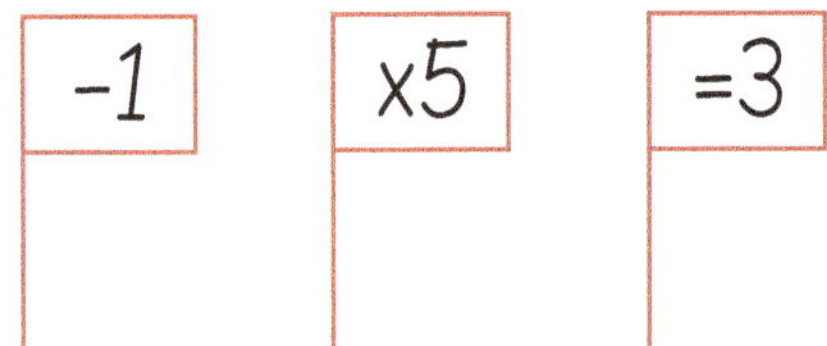

a. Complete all six flag diagrams below using the three flags.
b. Calculate the output for each flag diagram.

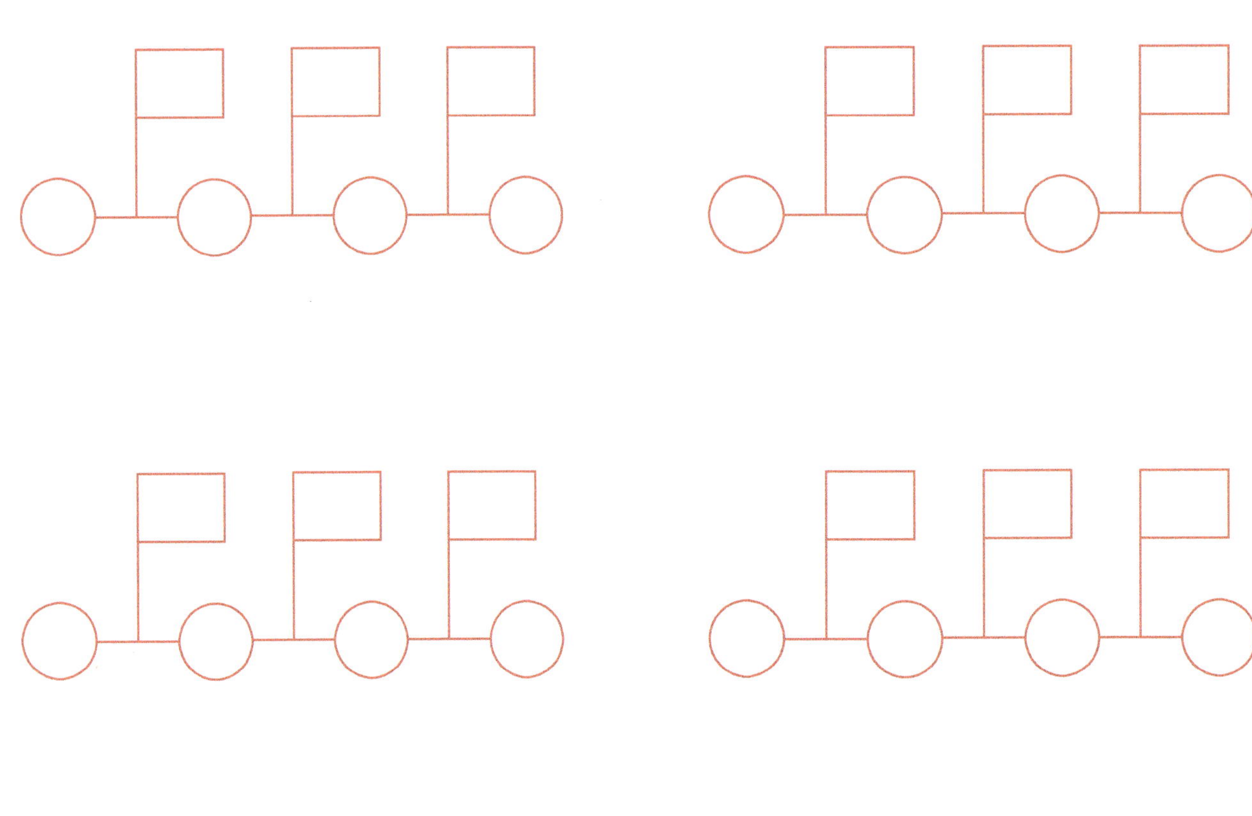

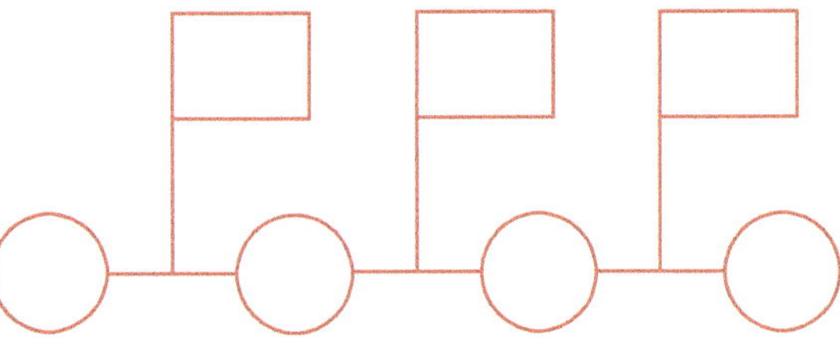

Flags
Level D

You have three flags available.

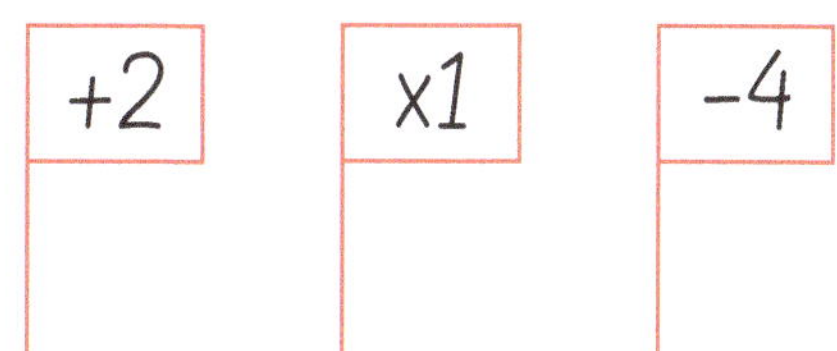

a. Complete all six flag diagrams below using the three flags.
b. Calculate the output for each flag diagram.

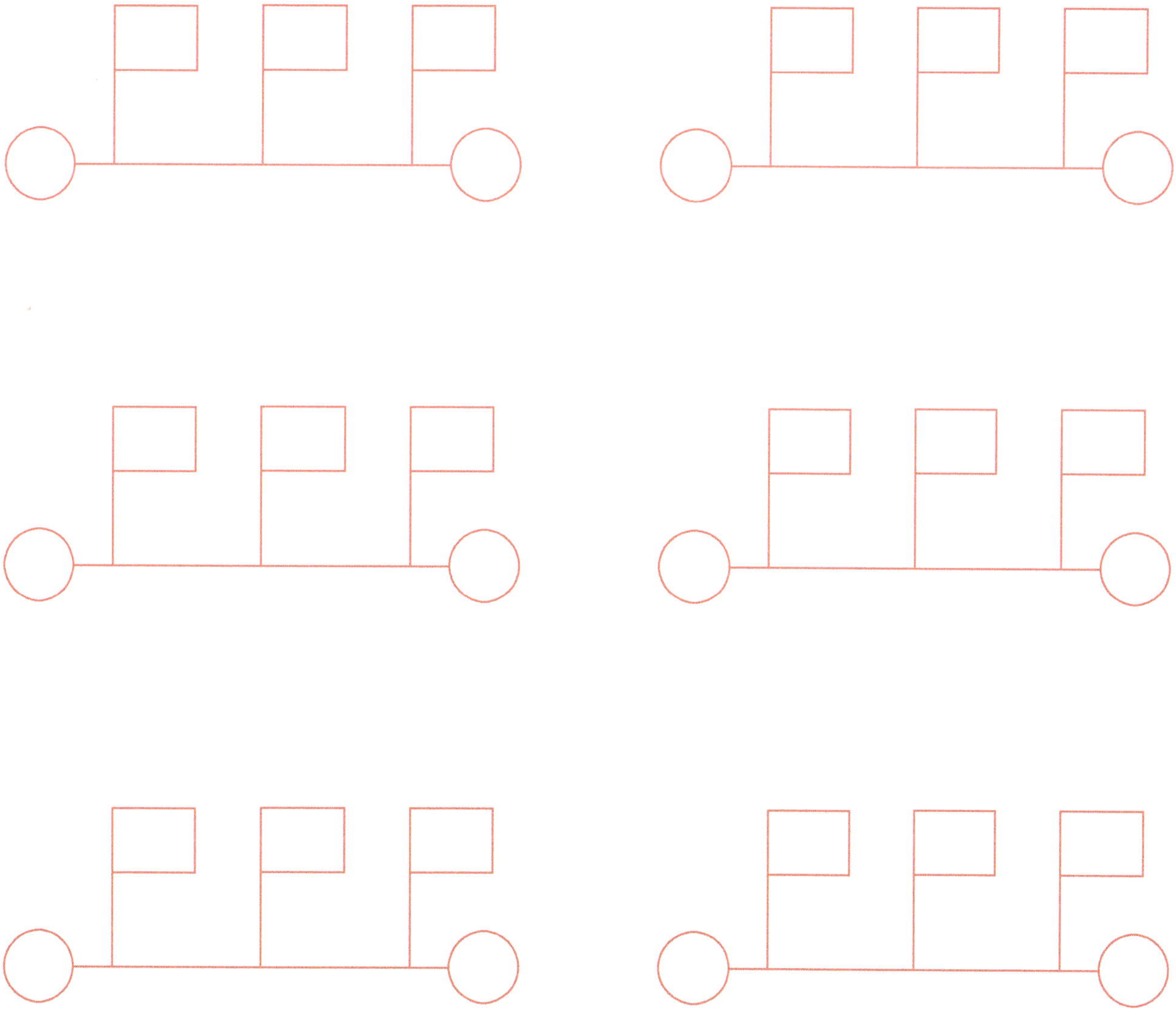

15

Flags
Level D

You have three flags available.

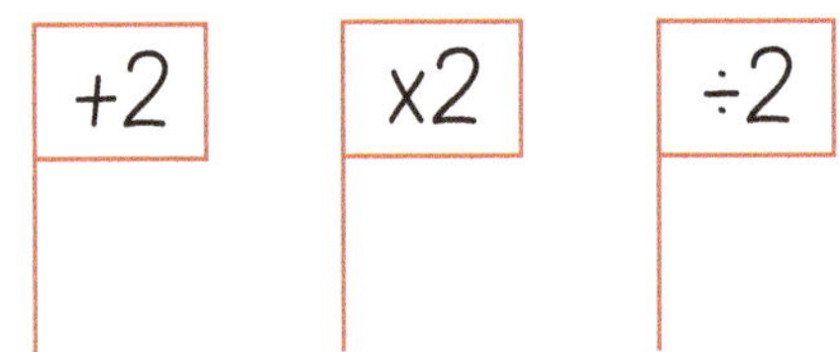

a. Complete all six flag diagrams below using the three flags.
b. Calculate the output for each flag diagram.

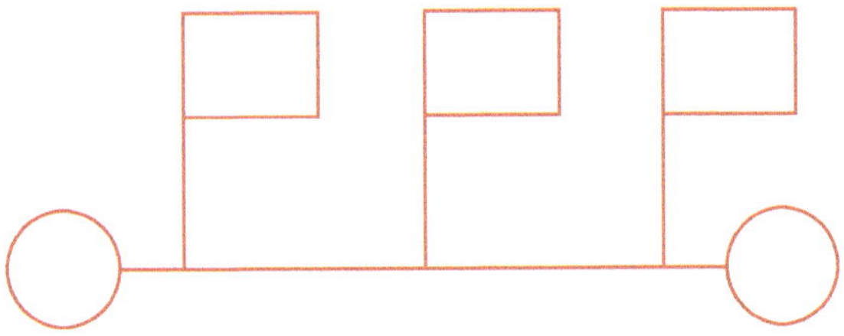

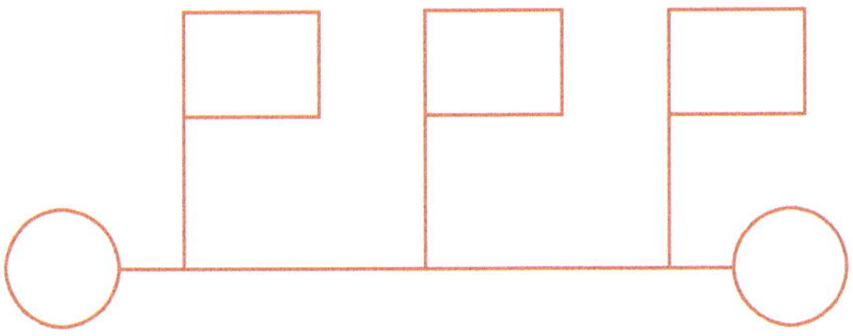

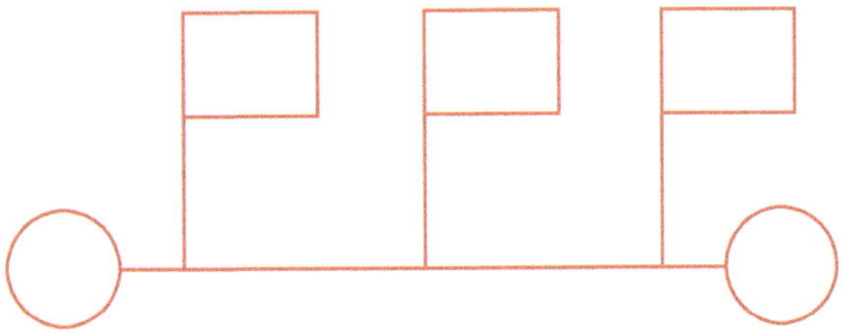

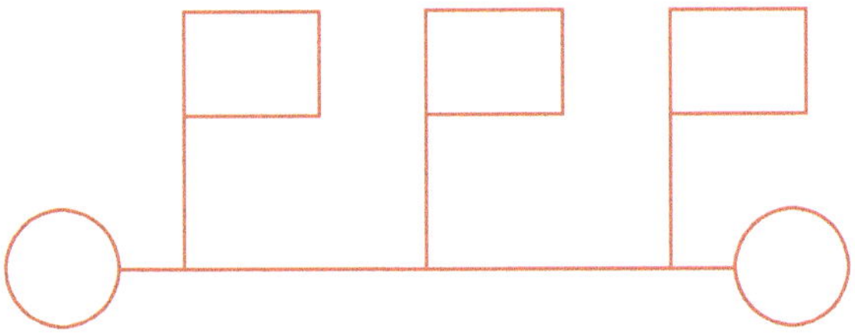

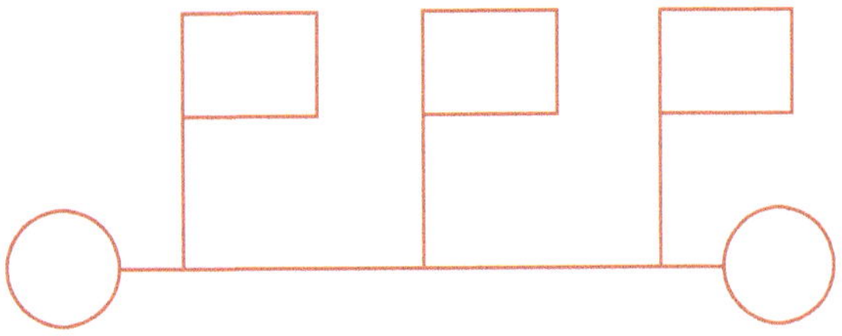

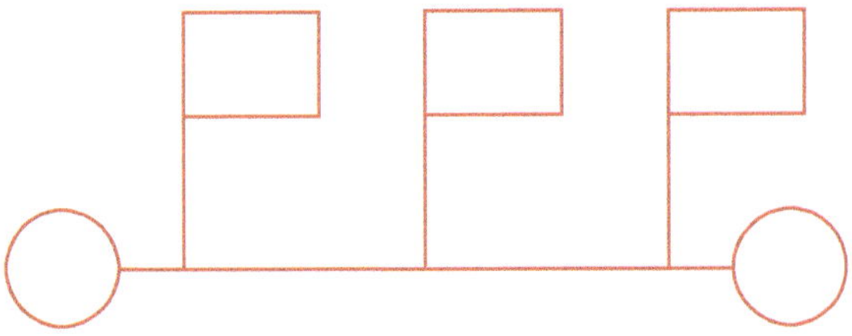

Flags
Level D

Problem 1

Tracy and Greg were playing a game.

Tracy said, "I chose a number, multiplied it by three and subtracted 2. My answer was 10. What number did I start with Greg?"

Greg thought for a moment and then drew the following flag diagram.

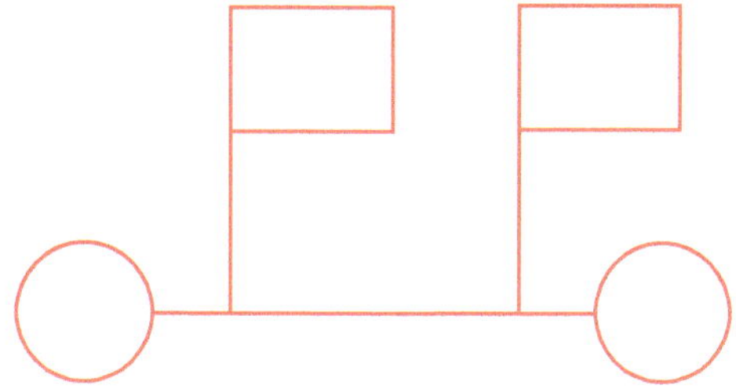

Use this flag diagram to work out the number.

Tracy started with ....

Problem 2

Greg had also made up a problem.

I chose a number, divide it by 2 then added 4. My answer was 10. What number did I start with?

Complete the flag diagram below then use it to find the number Greg started with.

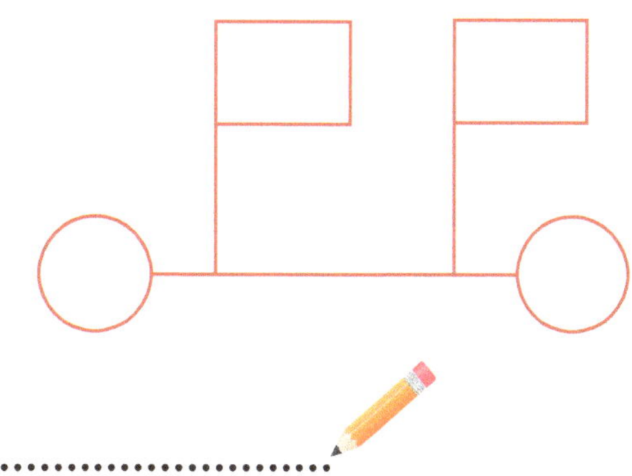

Tracy started with ....

Here are 4 more flag diagrams made up by Tracy and Greg

Write down the problem in words for each flag diagram. The first one has been started for you.

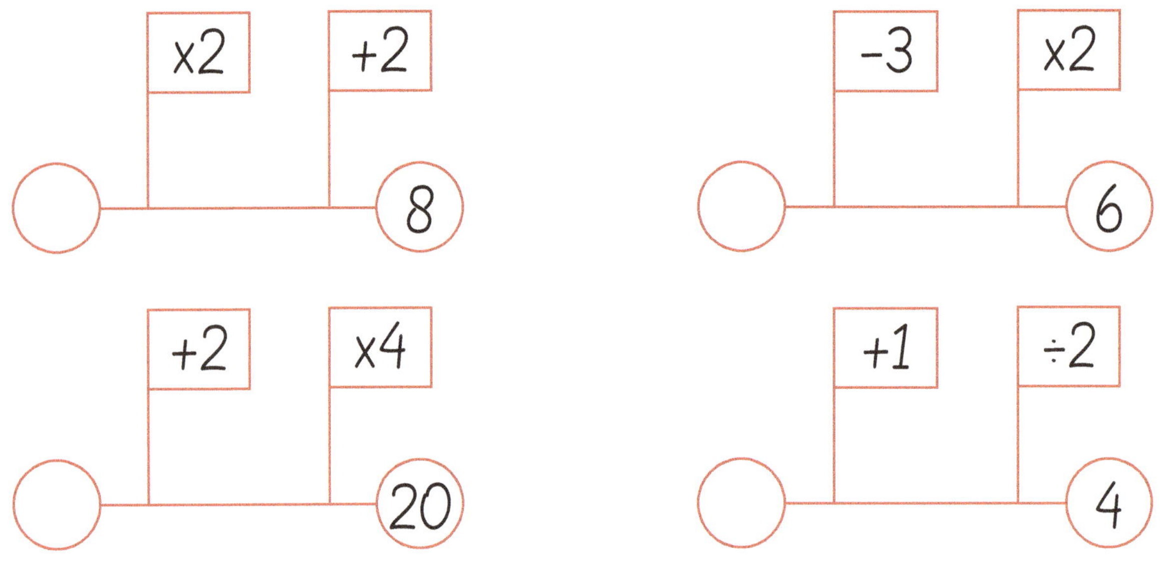

1. I chose a number, multiplied it by ....

2.

3.

4.

Flags
Level D   Activity 2

On your own make up **4** problems like Tracy and Greg did.

Write each problem in the space provided. The first one has been started for you.

Problem 1   **Problem 2**

I chose a number....................

Problem 3   **Problem 4**

Flags
Level D   Activity 3

Change sheets with your partner. Solve the problems you have been given using the flag diagrams provided.

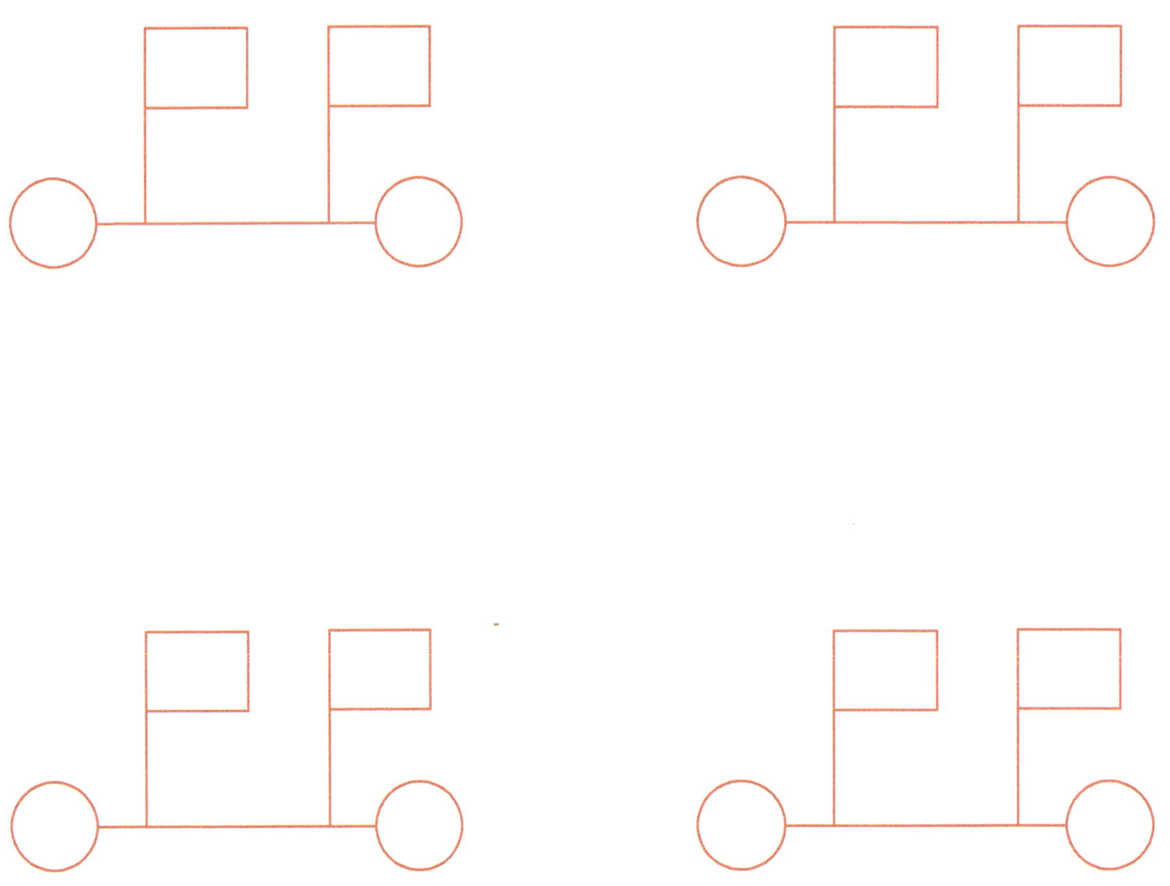

Check your partner's flag diagrams to see if they are correct.

Greg made up a new game using three flags

He told Tracy the answer was 18.

However, he did not tell Tracy the order in which the flags should be used.
Use the diagrams below to investigate what the input numbers could be.

The input numbers could be ....

Flags
Level D Problem 4

You have three flags available.

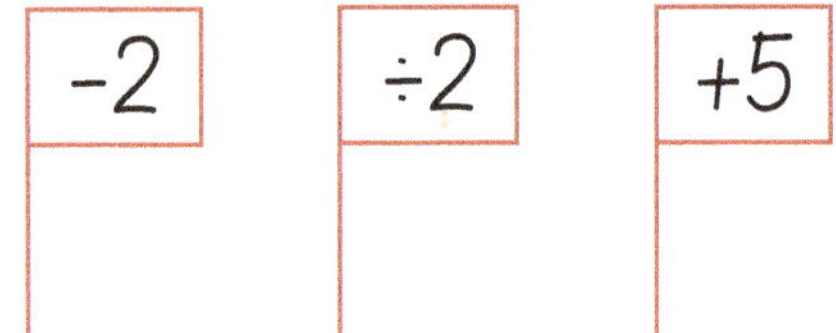

The output is 9. Using the flag diagrams below find as many input numbers as you can.

The input numbers could be ....

Here is a pattern with matches:

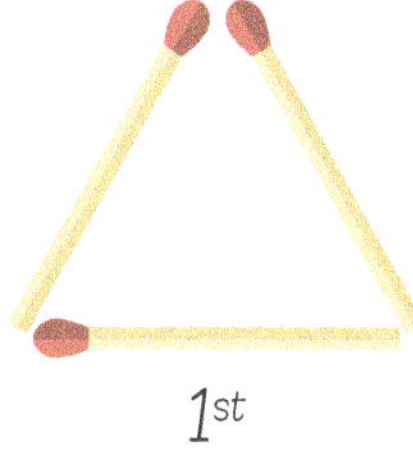 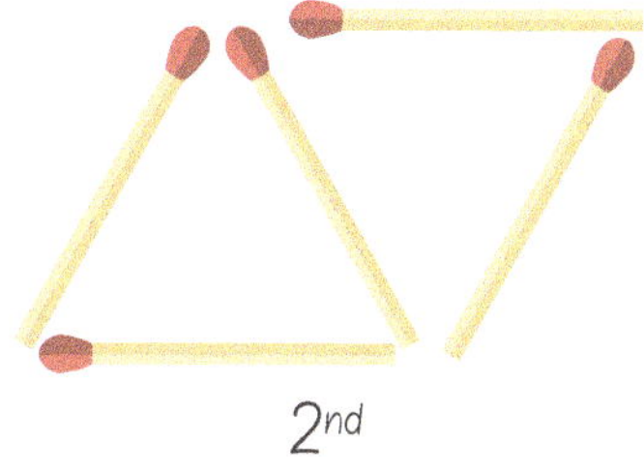 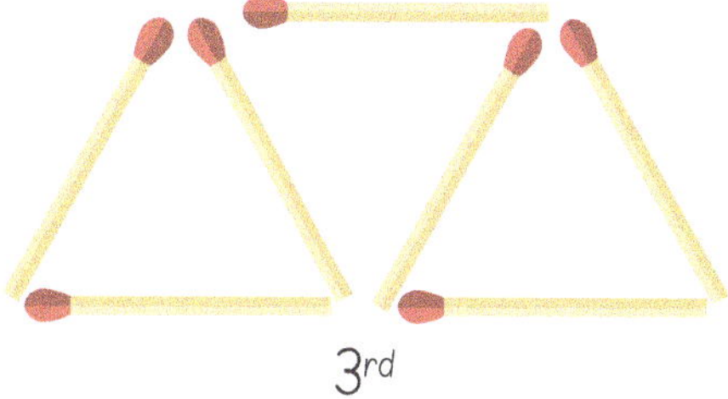

1st   *2nd*   *3rd*

Draw the matches to complete the 4ᵗʰ pattern.

Draw the matches to complete the 5ᵗʰ pattern.

Draw the matches to complete the 6ᵗʰ pattern.

Results can be recorded in the table below:

Number of triangles	Number of matches
1	3
2	
3	
4	
5	
6	

Complete the table of results. Can you spot a pattern?

Complete the sentence below:

The number of matches is ……. times the number of triangles plus ………..

15

Flags
Level D — Problems

Here is a pattern with matches:

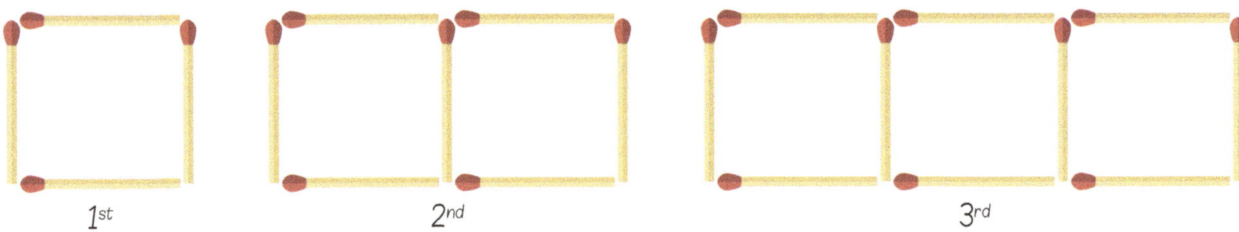

Draw the matches to complete the 4ᵗʰ pattern.

Draw the matches to complete the 5ᵗʰ pattern.

Draw the matches to complete the 6ᵗʰ pattern.

Results can be recorded in the table below:

Number of squares	Number of matches
1	4
2	
3	
4	
5	
6	

Complete the table of results. Can you spot a pattern?

Complete the sentence below:

The number of matches is .... times the number of squares plus ....

A fence made up of posts and rails is shown below:
copy and complete the fence:

How many rails are in the 1 gap i.e. between 2 posts? ....

How many rails are there in 2 gaps? ....

How many rails are there in 3 gaps? ....

The results can be shown in a table:

Number of gaps	Number of rails
1	3
2	
3	
4	
5	
6	

Complete the table above

Can you spot a pattern?
Write it down.

Extend the table for a longer fence up to 11 posts.

15

Notes

262

Notes

15

Unit 16

Flags and Plotting Points

Teacher Note

Functions and equations

Outcome: Number, money and measurement

Strand: Functions and equations

Target: Recognise and explain simple relationships between two sets of numbers or objects.

Target: D

Aims

So that pupils can understand the relationship between input and output numbers as represented by a number pair and by plotting the points on a graph.

Prior Knowledge

The pupil should be familiar with flag diagrams and have some knowledge of co-ordinates.

Plotting points
Level D

Here is a flag 

Here is a set of input numbers: 1, 2, 3, 4, 5

What are the output numbers?

Complete the flag diagrams below:

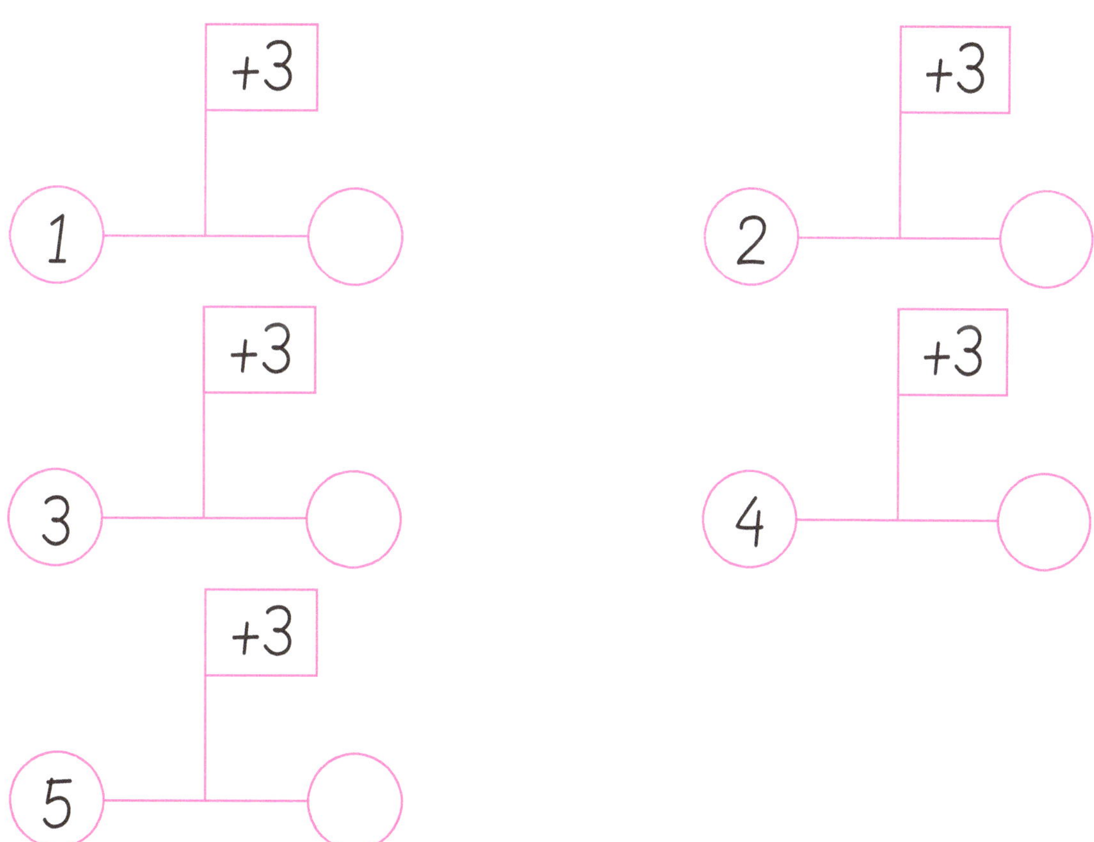

We can write the input and output for each flag as a number pair like this:

Complete: (1, 4), (2,.........), (3,.........), (4,.........), (5,.........),

16

Plotting points
Level D

These can be shown on a graph as:

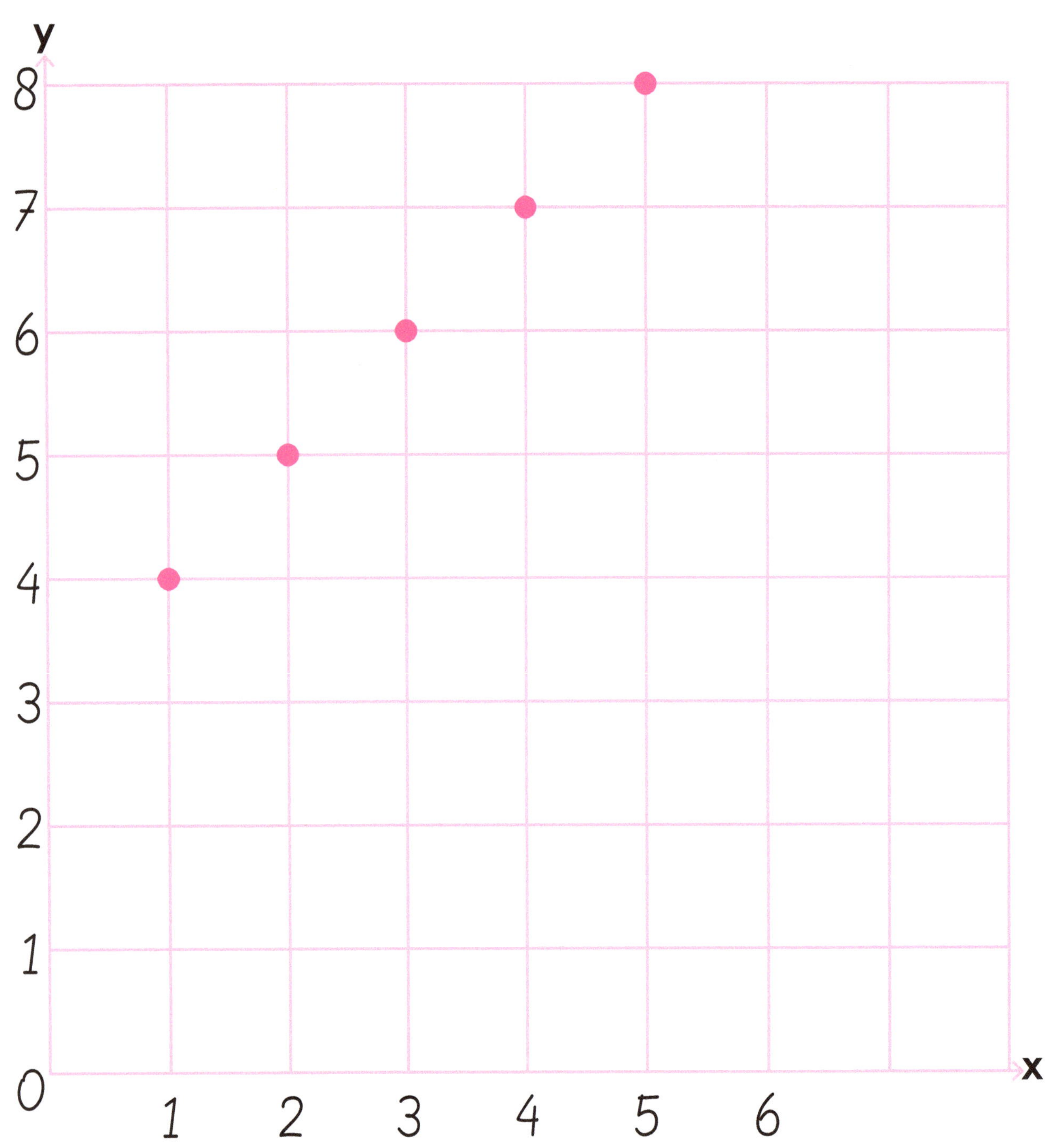

Plotting points
Level D

Here is a flag

+2

Here is a set of input numbers: 1, 2, 3, 4, 5

What are the output numbers?

Complete the flag diagrams below:

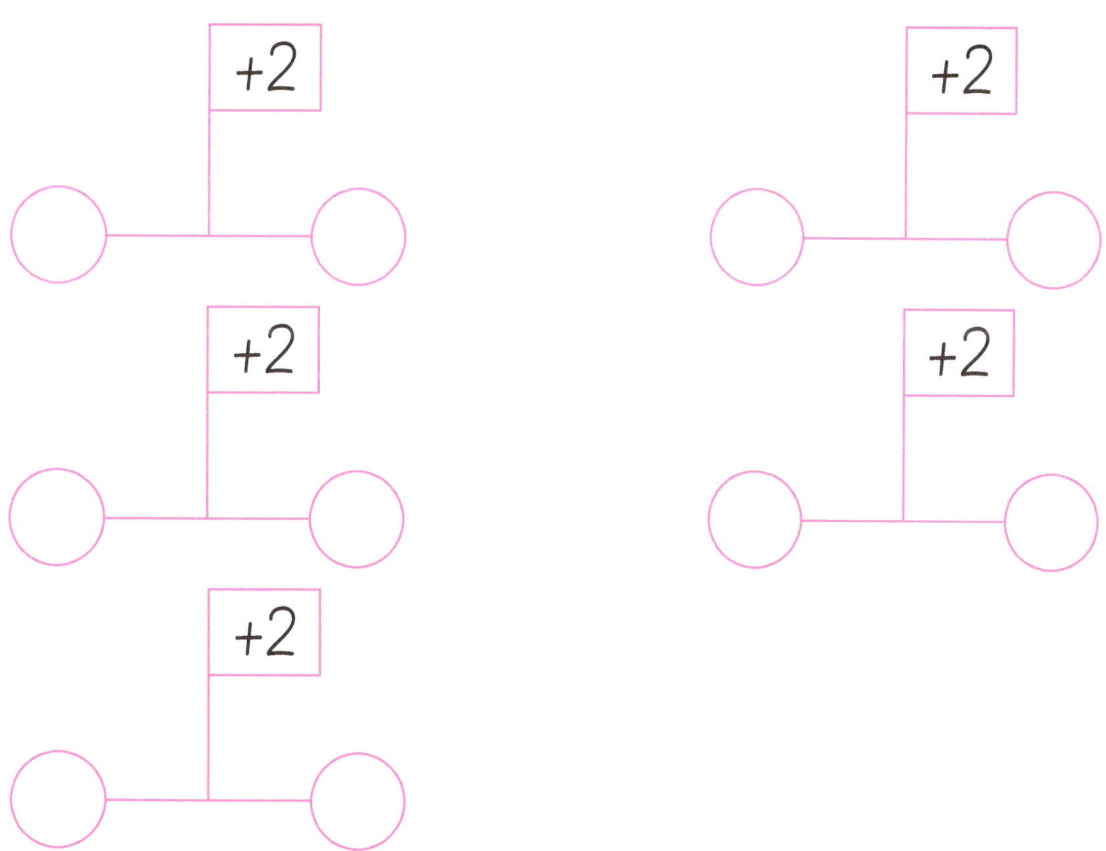

We can write the input and output for each flag as a number pair like this:

Complete: (1,.........), (2,.........), (3,.........), (4,.........), (5,.........),

16

Plotting points
Level D

Now show these points on a graph:

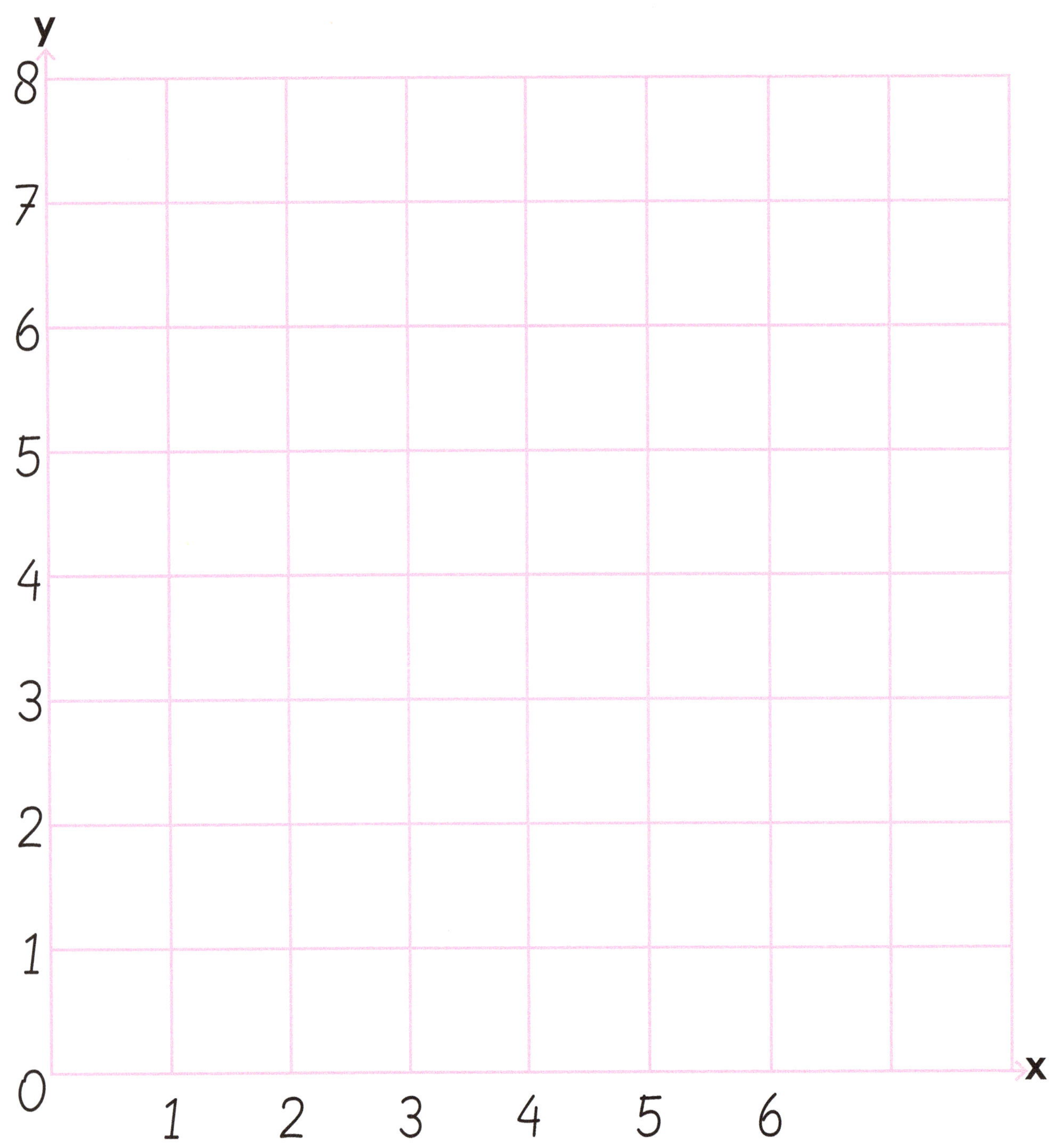

Plotting points
Level D

Using the blank flag diagrams and graph chart complete the process for the following flags:

1. **+4 and +5**

    with input numbers 1, 2, 3, 4, 5.

2. **-2, -3, -4, -5, -7**

    with input numbers 8, 9, 10, 11, 12.

3. **x3**

    with input numbers 1, 2, 3, 4, 5.

4. **÷2**

    with input numbers 2, 4, 6, 8, 10.

Plotting points
Level D

Here is a flag

Here is a set of input numbers: ....

What are the output numbers?
Complete the flag diagrams below:

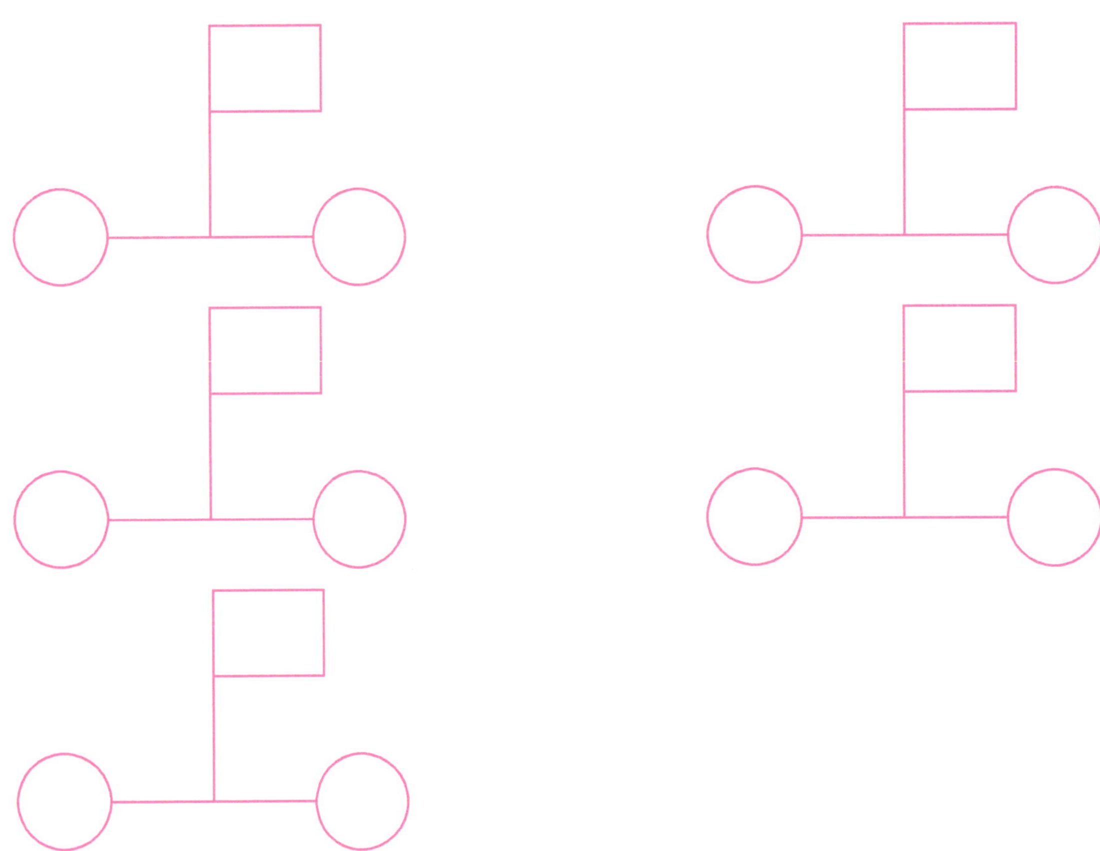

We can write the input and output numbers as number pairs:

(...............), (...............), (...............), (...............), (...............)

Plotting points
Level D

Now show these points on a graph:

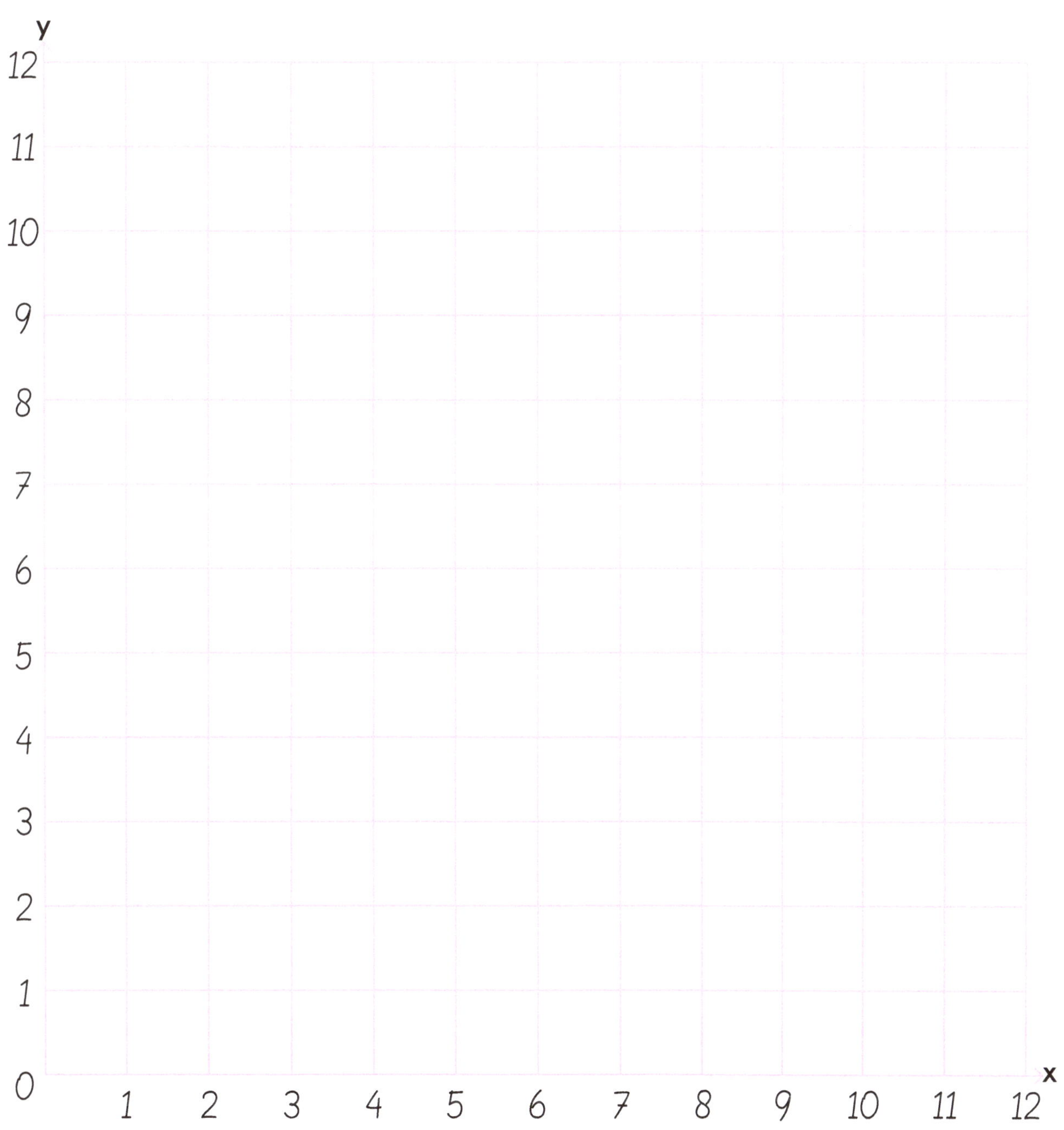

273

16

Plotting points
Level D

Here are two flags

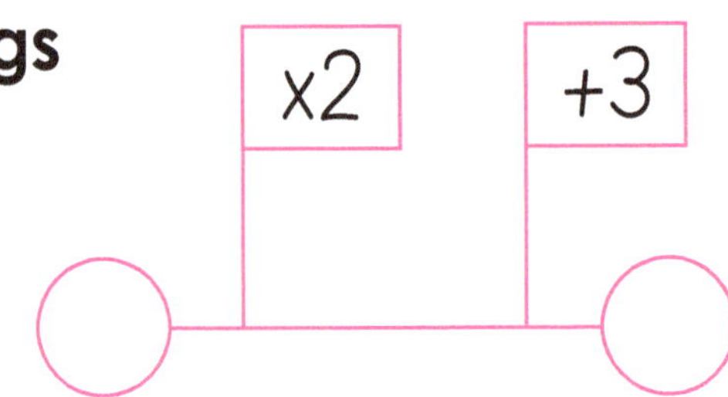

A.   **1.**   Choose 4 suitable input numbers and find the output for each.

       2.   Write the input and output values as co-ordinate pairs.

       3.   Draw axes and label them.

       4.   Plot the points on the graph.

B.   Follow the steps above for

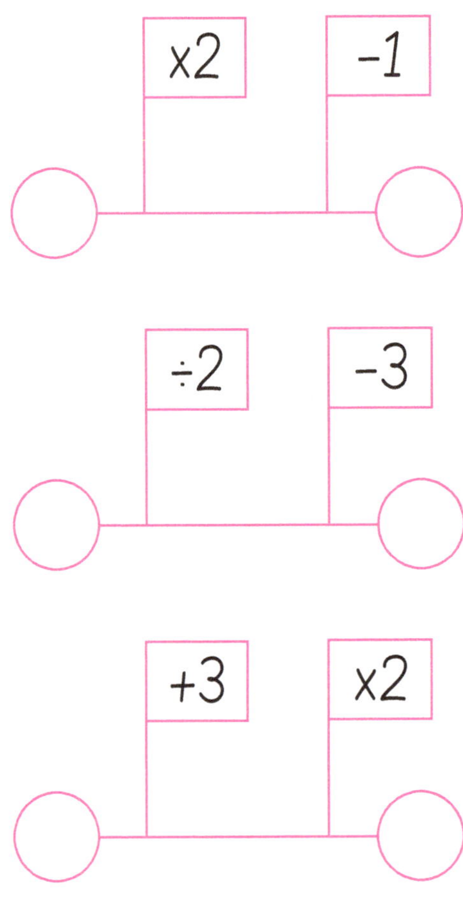

16

This can be repeated for x5 and x10 at level B

ALSO

for x6, x7, x8, and x9 at level C.

Notes

276

Notes